Foreword

Hi, my name is Raymond Palmer, I am a PhD Candidate and the author of The Anti-Aging Toolkit. I work in biotech as an engineer and longevity researcher. Through my work related research I have established a regime of longevity technologies which I have simply amalgamated and simplified for you, into the ultimate
"anti-aging toolkit".

I am your resident data miner!

How Should You Use This Book?
You must always speak with your health care provider before using or doing anything you read in this book.

If you want to lose weight, there is a section at the back just for you. If you are trim, then you can ignore that section. But know this, holding excessive weight is one of the top instigators of accelerated aging. I will teach you "the science" of weight loss, not just what to eat.

All the information in this book is based on studies by some of the world's most prestigious scientific, educational and research institutions.

I rely on peer reviewed studies which have been verified by very credible third parties.

I am a passionate believer and implementer of these anti-aging technologies myself.

And while it is still too early to tell if the studies will lengthen human life by any significant measure, the evidence (so far) indicates that it will.

You will see many references to the names and institutions of the discoveries allowing you to further research the claims I make.

A scientific reference will look something like this - (Palmer et al, 2013, AGELESS, Vol 35, 189-195)
And you may see those references in the body of the text or at the foot of the page.
In the reference shown, Palmer is the surname of the scientist, then the year he wrote the dissertation, the scientific journal followed by the volume of that journal and then page numbers. Pretty straight forward really, but it allows you to perform further research should you wish.

Sometimes you will see the letters 'aka', which means 'also known as'. Some compounds have more than one name, so I like to use aka between the names.

We have done our best to ensure accuracy, but by reading this book you agree that you accept that there may be errors beyond our control. So do your own research and speak to your healthcare professional before taking any advice from this book.

Note, I never rely on (and neither should you) information found on .net, .com or .org websites. All data is from .edu or .gov domains (because the standard is peer reviewed by other experts, therefore significantly higher).

Acknowledgements

Without the army of warring scientists out there fighting against aging and devising attack plans to push it back and hold our biological ground, this book would not have been possible. So thank you to all those extremely hard working scientists that made this knowledge possible.

This book could have been 500 pages, but it has been kept short and sweet so that you, the reader, can absorb the science quickly and put it to practical use.

Thank you to Leanne, Ronnie, Hunter, Sally, Yoda,
Jack, Mike, Lindsay, Margaret,
Dr Howard Chipman,
Professor Mauro Vaccarezza, Dr Sandra Kaufmann,
Mr Leech, many others,,,

1st Edition ver 2.3

www.ScienceOfAging.life

This book does not and should not constitute medical advice.

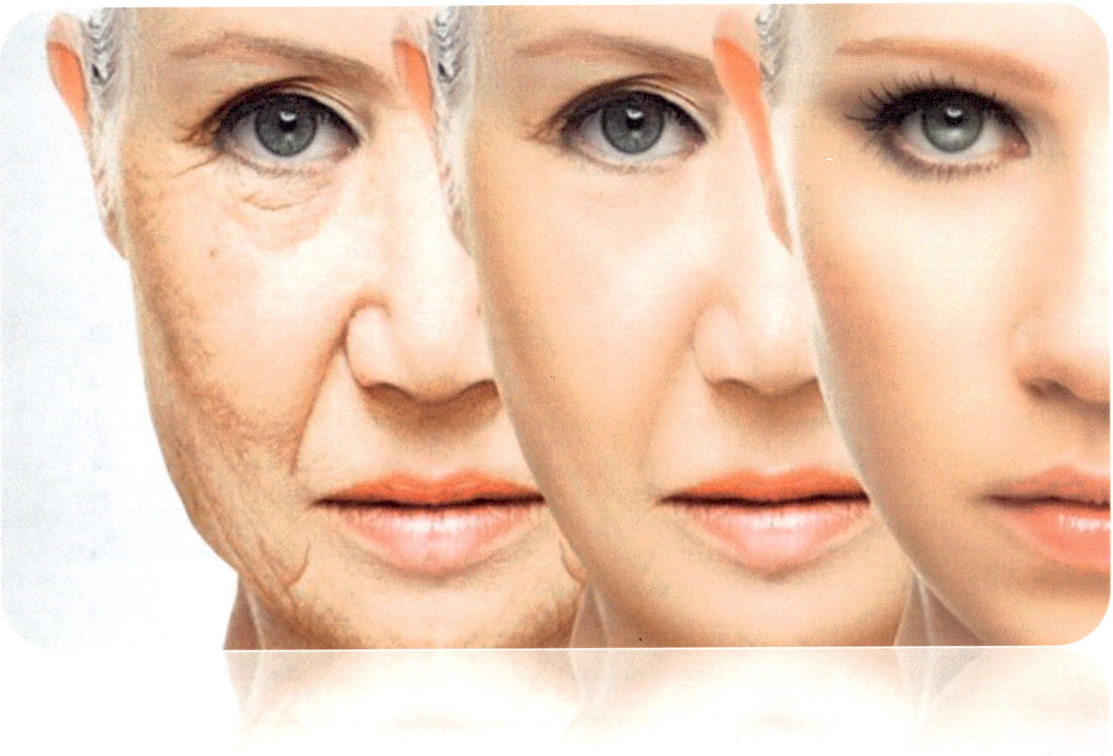

ScienceOfAging.life

Introduction

This book will start by listing some popular products that you can easily find online or at your local health food shop. Keep your eye on my social media accounts for the latest information.

The book will then move onto how some hormones are related to aging and how to control them.

We will finish with some information on how diet speeds up aging and how diet can also slow it down. Worth noting that food is medicine and food is poison. Also knowing when to eat and when to slow down your consumption also has a profound effect on your longevity (look for our fasting / calorie restriction section).

We will also discuss which exercises can slow down aging or speed it up. Finally we will discuss some fringe anti-aging technologies that are all in the pipeline with many already undergoing clinical trials.

Again, everything in this book can be verified by published scientific papers which you will find are regularly referred to.

This book is only a glimpse into the fascinating world of longevity technology and what science is currently undertaking. This book was once intended as a coffee table book, however to capture just a glimpse of this world, we've gone much farther but also kept the book easy to pick up and absorb. I have intentionally left out terminologies and in depth scientific discussion that the lay person may not be familiar with.

I have also left out much of the scientific principles and theory surrounding how some of the anti-aging technologies work. Those types of discussions are beyond the scope of this book and would require a completely different format and approach. This book is about practical science and is a 'how to' on lengthening your health span and life span.

The science is real and factual. Every case study cited in this book is backed up from peer reviewed medical journals or clinical trials.

Some of the scientists mentioned in this book are the anti-aging titans spearheading some of the latest research and pioneering the latest techniques across the world. If you are serious about longevity, then I bring their identities to your attention because they are worth keeping an eye on.

This is an updated edition, the most up to date version, however every year I may update and release a newer edition with the latest research so that you know what is working and what new discoveries have come to the fray. Further editions will contain most of what is already in this edition so do not expect an entirely new book; however, any new edition will always contain information worthy of another visit. This field is now delivering new science constantly.

The Anti-Aging Scoring System

I am going to bombard you with a ton of molecules, compounds and supplements. By the end of this book, your head may be spinning on which ones to prioritize. So, under each molecule / supplement section in this book, I offer a score out of ten, the higher the number, the better it appears to work for longevity.

You may be on a budget and only able to purchase the top three, but some of you may be able to purchase the top ten or fifteen, so this is why I felt a scoring system had to be used. This system will help you make good decisions and hopefully prevent you from wasting money.

Some molecules / compounds I recommend are still being argued about by other scientists on their efficacy and power over human aging. Some of these molecules have been left out after further investigation, but others, which teeter on the brink of offering something good have been included. So don't be surprised if you see a scientist somewhere arguing something different, because this industry is moving incredibly quickly, and longevity supplements are either being superseded or encouraged by new research weekly.

Note, I am a tough critic, so a six or seven in my scoring system is a good score.
An eight or nine? Wow!

So, use my scoring system as a guide to assist you with what to use. You can always shoot me questions on social media –

@RayDPalmer
YouTube @Science_Of_Aging

Life Span versus Health Span

What is interesting is that humans are living longer. The longer we live the more we age and the more age-related diseases we get.

So current government logic (and spending) is to fight these age related diseases with extremely costly and lengthy reverse engineering based research.

Ailments such as heart disease, cancer, arthritis and neurodegenerative diseases are in store for all of us, we know this, the government knows this, and in a crazy mindset of reverse engineering, 99% of all our research funding goes into curing, treating or managing these age related diseases, when the panacea to age related diseases is to simply find a way to slow down the onset of age in the first place.

Keeping people young and healthy is the common sense approach to beat age related disease (for any rational thinker).

Logic

Longevity technologies cannot currently offer you immortality (but extreme longevity is closer than you think), this book is not about giving you an infinite life span, it is about giving you a much longer health span to suit your longer life span.

Imagine living to 115 and being fit and robust. This is where current technology is taking us and it is no longer science fiction. There are now many products hitting the market which can tell your cells and your DNA to function as it did when you were young.

However these technologies are so new it is still too early to know if these are going to have an effect if you take them for 50 years, however the science so far is extremely promising.

Factor in that 50 years from now longevity technologies will be hundreds of times more advanced than now and you may not die at all. This is not science fiction, there are already clinical trials being conducted with technologies to do this as you will read shortly.

So if you can make it through the next decade or two, get ready to be living inside a science fiction movie, because truth is stranger than fiction.

If you think extreme longevity is reserved for those movies based far into the future, think again, because with cloning, parabiosis, telomere technology (all to be explained) and a range of other techniques, you may never run out of stem cells and reach your cellular limit again.

Get ready to be enlightened.

It is worth noting that before I take you on that journey of extreme longevity that most, if not all of our anti-aging toolkit will be naturally occurring compounds and techniques most of which are already hidden away in nature somewhere. The science already demonstrates that these compounds hold profound effects for those of you ready to use them.
Which brings me to another extremely important point; there are two types of people reading this book, either a reader in neurological stasis (a rigid non-evolving mindset), or the evolvers (those that change habits with new information).

Well for those of you that are set in your ways, refuse to evolve, disease and certain death awaits you (your welcome). For those of you that can evolve with new science, maybe, just maybe you can stay ahead of whichever biological decay comes your way. Chances are, if you are reading this book, you are an evolver, but even for an evolver change can be hard, so rewire those neural networks and live long and healthy.

You are about to realize that you are already living inside a science fiction movie, because if you think human life spans of one hundred and fifty years, or two hundred years are still hundreds of years away, call Hollywood and let them know that those movies set far into the future with today's human life spans are completely wrong.

Welcome to our near future.

Metformin
Anti-Aging Score 8.5/10

Metformin is a well-known prescription drug mostly prescribed for diabetics that you will need to chat with your Doctor about before you take. However metformin is currently being considered for approval by the all-powerful Food and Drug Administration as a drug that may extend life and also protect you from various diseases.

The most exciting thing about metformin is that it is a slightly modified compound that has been around since 1640 from the plant known as French Lilac or Goat's Rue. Interestingly Goat's Rue is a centuries old remedy to stop frequent urination, a warning sign of diabetes. Today a slightly modified version of this plant is the primary drug given to diabetics' worldwide as metformin.

Researchers studied diabetics on metformin and compared them to diabetics on other forms of medication and found something incredible. Those on metformin were healthier in many different ways. Apart from living longer than their counterparts on other medications, there were fewer incidents of cancer (as much as 25 to 40 percent), fewer incidents of cardiovascular problems and they were even less likely to suffer from neurological diseases such as Alzheimer's or dementia. Even when they did get cancer they still outlived other diabetics with cancer that were taking different diabetes medications.

With 50% of Americans either pre-diabetic or suffering from type-2 diabetes, one can only imagine how many lives metformin has already saved from other diseases that were unrelated to diabetes.

Another diabetes drug known as rapamycin was tested on the longevity of animals and even outperformed metformin, and there are many new "human" trials with rapamycin. We'll get to Rapamycin later in this book.

Metformin works by reducing the amount of glucose released from the liver, which lowers blood sugars. This in turn means the pancreas needs to make less insulin. Sugars in the blood also lead to glycation, which is an accelerated aging process in cells. Yes, sugar ages you faster, an entirely different field which we will briefly discuss shortly.

I certainly take Metformin as well. At first my Doctor was not buying into it, he would only prescribe metformin if I were pre-diabetic or worse, but I predicted this professionalism and took him some printouts from the FDA and other .gov websites. It did not take long to change his mind.

You can thank Doctor Nir Barzilai and his team from the Institute for Aging Research at the Albert Einstein College of Medicine for this ground breaking research.

So, if you are a healthy person not taking Metformin, it's possible an unhealthy diabetic on metformin has a chance to outlive you, so if you are looking for something that is available today whilst reading this book, you've just found one. Go speak with your Doctor, only he can advise you properly.

Some side notes about Metformin though. Metformin can block the benefits of exercise. So, try not to take it before, during or within four hours after your exercise routine, this will allow you to get the best of both worlds. Some days that I exercise I don't take Metformin at all. And days where I know I will be consuming more food than usual, social event, wedding, etc, I'll make sure I have them with me.

And note, that research is still coming in on Metformin, with some studies saying that no life extending event was seen, however we are talking about mouse studies here. That said, this industry is moving incredibly fast, I find myself writing about something one year, and then seeing it superseded the next, so stay tuned.

Because Metformin can assist your body in lowering blood glucose, it keeps you closer to a fasting state. And fasting also assists with regeneration and longevity, and we will of course perform a deep dive into fasting and regeneration later in this book. Fasting is the number one way to increase longevity, across all organisms we test it on, and Metformin is considered a "fasting mimetic", because it helps mimics fasting inside the body (more on that later).

I take Metformin because I want to blunt my sugar spikes. Sugars spikes can cause all types of issues, but for me personally, I want to slow down "Epigenetic Drift", which sugar spikes appear to speed up, and you will learn all about epigenetic drift later in this book as well. So, for me, Metformin continues to be part of my stack which I take carefully around my exercise regime.

Some take home messages for Metformin –

- People under 40 should not use it for longevity.
- Try not to use it before or after exercise (4 - 6 hours each side depending on brand).
- Metformin lessens muscle enlargement in exercising elderly. However, those that exercise and use Metformin had more benefits than exercise alone.
- Metformin can keep you closer to a fasting state.
- Do not use Metformin if you are taking Rapamycin and Vice Versa
- None of this is medical advice, so always chat with a doctor literate in anti-aging and longevity technologies first.

Ref : Front Genet. 2022 Sep 26;13:955835. doi:10.3389/fgene.2022.955835.
Ref : Cell Metab. 2020 Jul 7;32(1):15-30. doi: 10.1016
Ref: Aging Cell. 2019 Dec;18(6):e13039. doi: 10.1111/acel.13039

Sugars and Glycation

If diabetes was not enough to put you off sugars, then keep reading. Sugars react with cell surface fats and proteins in a process called glycation. The sugars interrupt or disrupt various signaling processes so that the cell cannot receive or send signals properly anymore. This can lead to improper cellular function, total lack of cell function or cellular death (ultimately disease).

That may not sound too bad, but as an example, if you are not repairing your skin properly, guess what is going to happen? Your skin is going to remain damaged and age faster. Skin elasticity will be disrupted resulting in a more wrinkled appearance. The damaged cells not working as they should in other skin conditions may also become more prevalent such as age spots, hyperpigmentation, sagging, inflammation, even tumors. All of those conditions could be preventable if your cellular mechanisms were functioning properly.

But it is not just our skin we should be worried about, everything from your liver, eyesight, heart and brain function all require cells working flawlessly and too many bad cells can result in all types of serious health problems including cancer. As we age, get older and continue to consume sugary food and drinks our bodies absorb these sugars that then cause reactions known as 'advanced glycation end products' or AGEs (which is a useful acronym).

Of course we need sugars for our body to have energy, but the food industry is competitive and manufacturers are sticking anything and everything into their foods to entice you, and as a result sugars are finding their way into our bodies excessively. In 2013 researchers from the Leiden University Medical Center (Noordam et al, 2013, AGE, Vol 35, 189-195) conducted a study on the blood of about 600 participants both men and women to measure their blood sugar levels and then asked another group to guess their age by looking at photos. The results were incredible; those that looked older had higher blood sugar levels. The math worked out that for every 1mmol per litre increase in blood sugar levels the age was thought to be 5 months older.

Remember - Advanced Glycation End Products = AGEs

Lucky for all of us we can control our blood sugar levels by eating healthy foods that are not laden with sugars. Eat low glycemic foods and for example, three meals a day four hours apart with no snacking in-between (you will learn more about diet and longevity later in book). Steer clear of burnt and blackened meat. Apart from containing cancer causing molecules known at polycyclic aromatic hydrocarbons (PAH's), they contain those AGEs (mentioned earlier). Cook meats slow and low (preferably do not eat meat if you are serious about longevity).

If you have a few extra pounds try and lose them because larger people can induce the formation of AGEs. However building muscle is great as muscles require glucose resulting in cleaning up those AGEs. Antioxidants also protect cells from glycation in the first place so consume plenty of fruit and veggies. It's the same old advice we've heard time and time again really, exercise and maintain a healthy diet. Red meat, processed foods and cheese contain a lot of AGEs. Vegetable fats contain fewer AGEs than animal fats. Concentrated fatty dairy foods such as creams and butter can also contain a lot of AGEs (always select low fat dairy products or even better, non-dairy substitutes). Browning or burning foods such as toast or potatoes also increases the amount of AGEs in these products. Sugars come in the form of rice, pasta, bread and potatoes and are close to 100% sugar once broken down in the body. Keep this in mind. Plants based diets are best.

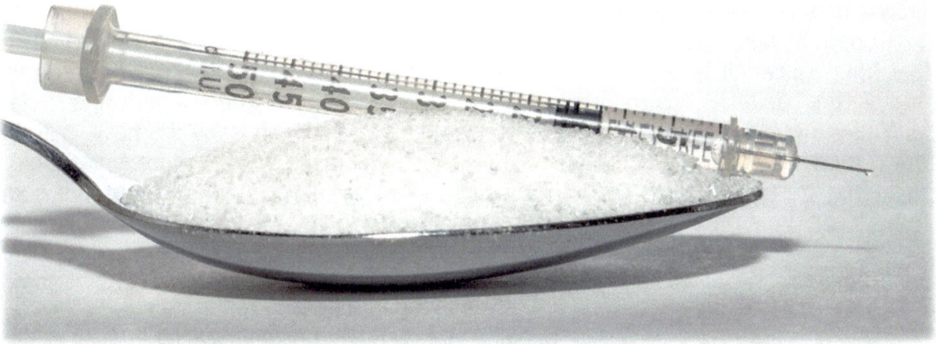

Are you addicted to this poison? Feel the need to add sugar or artificial sweeteners to your foods? Cannot handle popcorn without something sweet on it? Need to dip strawberries in chocolate first? Your taste buds will shift, stop using it, it will kill you. You will learn to find that sickly sweet taste of sugar quite disgusting over time.

Telomerase

Anti-Aging Score TBA

Our DNA's double helix strand that contains all our genes is wrapped tightly around 46 chromosomes. The ends of those chromosomes are known as telomeres. Imagine the flugelbinder on the end of a shoe lace; the telomeres are similar to those. Every time our cells divide our telomeres get a little shorter. The shorter they are the older (slower and poorer) our cells function until the telomeres become so short that the cells can no longer divide (the Hayflick limit) and either die or enter into a senescence state (a zombie like state which will be discussed later in this book).

As an example, we lose about 30,000 to 40,000 skin cells every minute!

It is almost like the telomeres are an instruction book for a cell on how to function. So how do we keep our telomeres younger? Is there a way to make them longer to induce a youthful cellular state? You bet there is. Telomerase is an enzyme that keeps telomeres long allowing cells to continue to divide.

- An enzyme is simply a protein that acts as a catalyst to induce and or speed up a chemical reaction inside a cell.

Regenerative cells in particular have high levels of telomerase to ensure the human body can continue to repair and replenish. On that note, telomerase could be considered an anti-aging enzyme. The function of telomeres and telomerase in aging were discovered in the 1980s and 1990s by the scientists Elizabeth Blackburn, Jack Szostak and Carol Greider who mutually shared in the 2009 Nobel

Prize in Physiology or Medicine for "the discovery of how chromosomes are protected by telomeres and the enzyme telomerase".

A study by Willeit et al took white blood cells from 787 candidates and followed them for 10 years. The study showed that telomere lengths and cancer rates were inversely proportional. The longer the telomeres, the less cancer, and the same was true for the shorter your telomeres the more cancer.

But do not overdose on telomerase just yet. For cancer cells to be immortal (which they are) they rely on telomerase in order to keep dividing, and low levels of telomerase can be just as bad for the healthy cells that fight cancer. It is almost like we need a perfect level for anti-aging, not too much maybe, but certainly not too little. Environment can also be a factor in telomere length and it starts from childhood. Young people that experience a lot of stress and injury in early life have cells that have divided much more by adulthood, shortening their telomeres making them more susceptible to disease. So remember, try not to cause children unnecessary stress if you want them to live a long and healthy life (children are a product of their environment).

Exercise is one of the best ways to boost telomerase naturally. Exercise clearly boosts telomere length, and as we saw in the study above, the shorter your telomeres are, the more cancer you may develop. This finding is consistent with other findings that those who exercise have less disease. Unfortunately, I have not seen any direct evidence or human trial data that the enzyme telomerase can be ingested and then find its way into the cell, and then further into the nucleus where the chromosomes (telomeres) reside. So sadly, getting telomerase into the nucleus from the outside is very hard, the cellular machinery to do that (cell surface transporters) don't seem to exist for this. However, other alternatives do exist. Cistanche tea which contains acteoside has shown increased telomerase activity, with the following list showing an increase in telomerase activity as well, but only in cells set in laboratory conditions or "in vitro", and not in human trials yet; Astragalus, Ginsenoside Rg1, Allicin (found in garlic or supplements), Withanolide from Ashwaganda and Basil Oil. Most of those can be easily found in supplement form with an online search. I cannot recommend dosage or brands, as there is a lack of clinical data to go off.

Your other options for longer telomeres are gene editing, where you can get a second telomerase gene "knocked in" with the idea of doubling your natural telomerase levels. This is already available, though the downside is the cost and invasiveness of having your genes changed. It is worth pointing out though, that gene editing for longevity is already here, I have colleagues that have undergone gene editing for longevity already, and I will also undergo a similar therapy in the very near future. Even though this sounds like the stuff of science fiction (just like Botox clinics were regarded a couple of decades ago), expect gene editing clinics to open in the coming years for health and longevity. Of course, there will be resistance to this, there always is with new fields of science, but what's worse? The onset of disease or keeping people healthy? I should note that these are somatic cells being edited, and these edits do not pass down to the offspring.

Other therapies are also being developed to extend telomeres, so watch this space and keep an eye on my social media (I try and update it when I can).

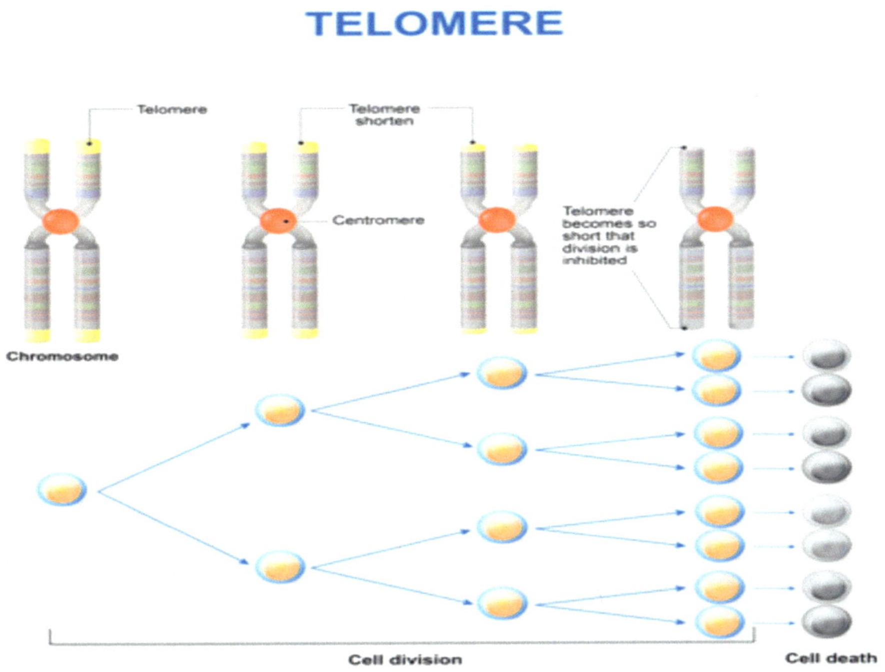

Note how the yellow end caps (telomeres) get shorter as cells divide? This divisional limit equates to old age, where you can no longer repair muscle, skin, bone, etc, resulting in you living out the rest of your life with dysfunctional cells making you slow and frail. But do not stress, this book will show you how to fend off this divisional limit and live much farther into the future.

Fanny, C, F. et al, Neuro Signals. 2014, 22, 52-63
Willeit, J. et al. JAMA, 2010 Jul 7;304(1):69-75
Yu-Sheng, C. et al, J Microbil Mol Biol Rev. 2002 Sep, 66, 407-4
Plant J. Med. Aromat. Plants. 2016;5:1–6
Kumar N., et al. Int. J. Curr. Microbiol. Appl. Sci. 2016;5:713–719.
Liu L.M. Chin. J. Hosp. Pharm. 2008;20:1767–1770

NAD - Nicotinamide Adenine Dinucleotide

Anti-Aging Score 8.5/10

When we are young our abundant energy comes from a powerhouse that resides inside our cells called mitochondria. As we age our mitochondria (the battery packs of our cells) produce less energy inside the cells making exercise and other humanly functions an effort. But why? Because NAD levels begin to drop, so like a game of dominos so do our energy levels and cellular function. However, boosting your NAD levels and keeping cellular function where it needs to be is easy with supplements that are designed to raise your NAD levels.

These two precursors induce NAD production inside the cells. Some precursors that will raise NAD levels are Nicotinamide, Niacinamide, Dihydronicotinamide Riboside (DNR), Nicotinamide Mononucleotide (NMN), Nicotinamide Riboside (NR). You should focus on the last two.

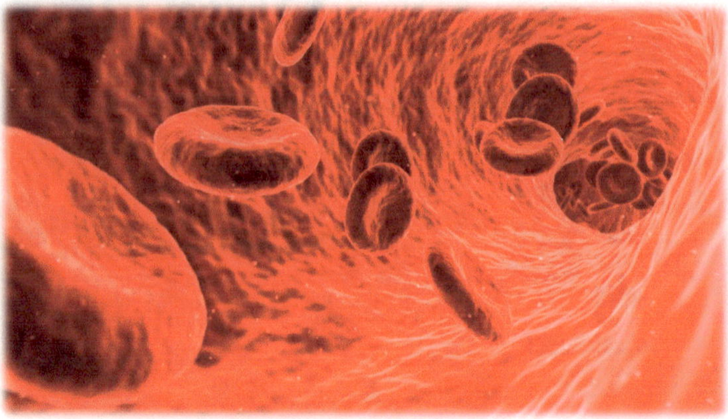

The NAD story is so much more than just your energy levels though, when NAD levels drop cells cannot function properly, and at this point disease can take over. Observe how an older person is frail and struggles to move freely, this is nothing more than the cells longer working as they should, unable to repair damaged tissue, joints and muscles. Glucose metabolism, cardiovascular function, stem cell health and neural function are all regulated by NAD levels. Professor David Sinclair from Harvard Medical School has shown overwhelmingly how important this coenzyme is in restoring communication between the cell nucleus and mitochondria (without that communication aging accelerates).

- A coenzyme is simply a molecule that assists an enzyme to carry out a reaction.

If NAD is 100% when we are born, by the time we are 20 it is at 50%. NAD is critical for cellular communication, meaning if cells cannot speak to each other, they cannot repair and function as intended. This is why 60 year olds take much longer to heal cuts and bruises than 20 year olds. You get the picture. So an over the counter NAD stimulator such as nicotinamide or nicotinamide mononucleotide (NMN) can be very handy to steer off an age related disease and or lifestyle. Even lethargy may be attributed to low NAD levels. Keep those cells supercharged with energy and your body will in turn bound with energy because food and exercise can only do so much.

Imagine doing some push-ups; your brain sends a message to your muscles, your muscles move, your heart pumps more blood to fuel those muscles and so on. Not only is a vast amount of biochemical communication going on, but a vast amount of biochemistry must instantly take place.
This communication and cellular function is all powered by NAD, so now that are you getting the picture of how important NAD is when it comes to aging, it is time to take NAD very seriously if you consider health and longevity important.

Your body may repair slowly, your body may feel old in the mornings and even during the evenings, you may always feel tired, or you may simply have the onset of age related disease (arthritis, osteoporosis, cancer, heart disease). It is very possible that all of those symptoms are caused by low levels of NAD. As a car engine begins to run poorly without regular servicing and good quality engine oil, so too will your cellular engines if you are not refueling the cells and mitochondria with NAD.

NAD is one of the most important coenzymes in your anti-aging took kit if you are serious about an increased health span and life span.

Some final words about the importance of NAD. NAD performs other functions are paramount to your anti-aging tool kit.

Lack of NAD is a major factor in cancers, as NAD is responsible for ensuring mutated genes do not get copied. NAD can plug holes in genes and also prevent faulty genomes from being copied from one cell to the other, yet there is even more to this miracle coenzyme (my paper and others below).

Ref: Palmer RD et al Precursor comparisons for the upregulation of NAD. Aging Med 2021 Aug 4;4(3):214-220
Ref: Suave, A. 2008, Pharma & Experi Thera, 324, 883-893
Ref: Imai, S. 2014, Trends Cell Biol, 24, 464-471
Ref: Sinclair, D. 2013, Cell, 155, 1624-1638
Ref: Bondowski, M & Sinclair, D. 2016, Molec Cell Biology 17, 679-690

Vascular Aging

Vascular aging is when blood vessels get old and begin to shrink cutting off the blood supply to organs such as the brain, skin and even muscles. As an example, muscles can shrink, and it's a reason why old people are often skinny and frail. Muscles require large amounts of oxygen to function, and if your capillaries are slowly disappearing and cutting off blood which carries that oxygen, then your muscles begin to shrink. It is the same with all your other organs such as the brain; imagine what happens to the brain as the blood and oxygen supply slowly withers away during old age? Your brain is going to shrink as well possibly leading onto forgetfulness, clumsiness and worse, neurological diseases that induce early mortality.

NAD (which we discussed on the previous page) is also paramount in blood flow. Age related muscle loss is known as sarcopenia and is an indicator of low NAD levels. Erectile dysfunction is also a good indicator or vascular constriction. NAD has been shown to improve blood flow and the creation of new blood vessels to various tissues. Activities such as exercise hold many benefits and cannot be underrated, but even exercise is at the mercy of your body's biochemistry and cannot completely prevent the onset of muscle loss as we age.

However, science delivers profoundly here and can reverse the loss of blood vessels throughout the body with NAD.

Another player here (which we will learn more about later) is a protein known as Sirt1. Sirt1 is elevated by NAD and is what allows muscles to communicate with blood vessels telling them where to grow and deliver nutrients and oxygen. Both NAD and Sirt1 decline as we age hampering the conversation between muscle and blood vessel growth (angiogenesis).

However, Harvard Medical School has shown that NMN which includes NAD can not only reverse this muscle loss, but increase fitness capacity by up to 80% in animal models.

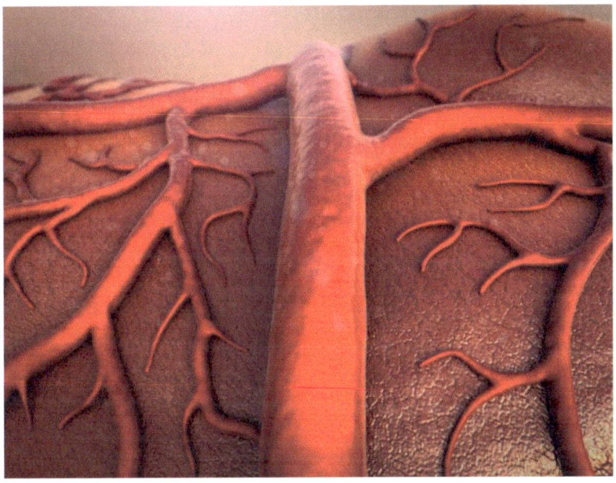

NAD is arguably one of the most important longevity enzymes discussed in this book. I certainly use NMN, NR and NAD orally, intravenously and also through the intranasal pathway daily (more about absorption techniques later). The principle of muscle loss through vascular aging can be and should be applied to all parts of the body including all major organs including brain and skin.

Vitamin K2

Anti-Aging Score 7/10

I rarely promote vitamins for longevity, but vitamin K2 has some interesting qualities. As we age, there can be calcification in connective tissues which are implicated in pathologies such as cardiovascular disease and osteoarthritis, which are highly common amongst the aging population. Vitamin K appears to inhibit vascular calcification by assisting a specific group of proteins to ensure that calcium is delivered to the correct places such as bone and not our arteries. Note that vitamin K1 resides mostly in the liver, whereas vitamin K2 is system wide and circulates around much of the body. This means that you could consider K2 the calcium vacuum cleaner for the inside of the arteries. This does not mean eat greasy food and simply supplement with K2, you will always need to take care of your diet, but it is great to know that calcification has solutions. And just to complicate things further, there are numerous types of K2 on the market, though you should target the type called menaquinone MK-7. Mk-7 is also found in a Japanese dish called natto, so worth keeping that in mind next time you are choosing sushi or noodles.

From here, I will start referring to K2 as MK-7. So what else is it good for? Mk-7 has been implicated in lowering inflammatory markers, offers protection against free radicals, controls calcium delivery to correct tissues, reduces arterial stiffening, effective in the prevention of osteoporosis, but here is the doozy, it is 10 to 100 times more powerful than any other known free radical scavenger! Do you smoke? Probably not if you are reading this, but if you live a life where free radicals are running rampant in your biology for whatever reason, then this is a compound for you. And note one final study showed that by taking just 375 micrograms (not milligrams) for one year prevented any loss in bone density, which is huge news for those that may have osteoporosis run in their family (and there was also a 60% reduction in fractures in older people). And some further claims found in the literature was that just 10 micrograms per day reduced diabetic risk of up to 7%, and 360 micrograms daily of Mk-7 reduced calcification inside renal (kidney) arteries, and an in vitro study (studies performed in the lab) was that K2 prevented growth and metastasis of various cancer cell lines.

Do not let your local health food shop sell you another type of K2, those other types either do not get absorbed well, do not stay in the blood for as long, or simply do not perform as well as MK-7 (I have done the research for you, so do not waste your money being sold junk). Recommended dose? Some countries say 50 micrograms, other countries recommend 600 micrograms daily. Me personally? With heart disease and osteoporosis in my family coupled against my plant-based diet and weight training, I will aim for 300 micrograms per day (and I will remind you again, that is not milligrams but micrograms when dosing). Always chat with your doctor first, especially if you use Warfarin.

Vitamin K: Double Bonds beyond Coagulation Insights into Differences between Vitamin K1 and K2 in Health and Disease. Int J Mol Sci. 2019 Feb 19;20(4):896

Vitamins K1 and K2: The Emerging Group of Vitamins Required for Human Health. J Nutr Metab. 2017;2017

Vitamin K as a Powerful Micronutrient in Aging and Age-Related Diseases: Pros and Cons from Clinical Studies. Int J Mol Sci. 2019 Aug 25;20(17):4150.

Vitamin K2 (menaquinone-7) prevents age-related deterioration of trabecular bone microarchitecture at the tibia in postmenopausal women. Eur J Endocrinol. 2016 Dec;175(6):541-549

Urolithin A

Anti-Aging Score 6/10

Another standout that has made my anti-aging toolkit is Urolithin A (UA), which is created when gut bacteria take the ellagitannins found in pomegranates and convert them into UA. Urolithin A induces a process called mitophagy inside cells, where the cells batteries (mitochondria) are broken down and then recycled back into fully functioning mitochondria. This is important because if your mitochondria are not performing properly, then cellular energy will be low and many of the functions that the cell needs to perform will be inhibited, which can and does lead to disease. Recycling the components of mitochondria back into new mitochondria ensures cellular longevity and youthful cell function.

It is still early days, but when given to worms it increased their energy levels by 50%, to rodents UA increased energy by 40%, and human clinical trials are currently underway. An interesting paper which I will reference below shows that UA suppresses amyloid plaque buildup which may prevent neurological decline. It can do this by suppressing high blood glucose which is also known to be a driver of amyloid plaque accumulation (among other negative aspects). If you do not already know, Alzheimer patients all have high levels of amyloid plaques in their brains, so keeping those plaques down is extremely important as we age. Remember, you need gut bacteria to produce UA (unless you are supplementing directly with it), so caution when using wide spectrum antibiotics, because antibiotics kill much of the good bacteria you need to be healthy (more on that later).

Inflammatory bowel disease is also positively influenced by eating several natural plant foods such as walnuts, berries or pomegranate which all contain ellagitannins and ellagic acid. It is believed that the UA that is produced from the bacteria acts as a potent anti-inflammatory and antioxidant agent. However even though UA has other benefits, it is mitophagy that makes this compound part of my toolkit. If you choose to use a supplement and cannot find UA directly, then try finding an ellagitannin or ellagic acid supplement (and try and steer clear of antibiotics if you can).

Urolithin A suppresses high glucose-induced neuronal amyloidogenesis by modulating TGM2-dependent ER-mitochondria contacts and calcium homeostasis. Cell Death Differ. 2021 Jan;28(1):184-202

Spermidine

Anti-Aging Score 6/10

Spermidine is a polyamine which activates our cellular recycling system known as autophagy (more on autophagy later in this book). I will not bore you with the biochemical pathway that spermidine works on either, but in humans, when spermidine is taken at elevated levels, researchers observed reduced blood pressure, lower incidence of cardiovascular disease and there is patchy evidence coming in that it may delay neurodegeneration in both mice and non-mammalian models. Two population studies confirmed that spermidine reduces cardiovascular and cancer mortality in humans. Biomarkers such as age, weight, lifestyle and even socioeconomic status were all taken into account, and those using spermidine appeared to have reduced mortality.

Spermidine has also been implicated in increased telomere length and decreased arterial aging. You do not need my level of knowledge to know that the increased telomere length and increased autophagy are the main drivers for the longevity magic here. Spermidine is found in soybean (cooked), green peas, pear, lentil soup, mushrooms, red beans, broccoli, cheddar cheese and many more foods, otherwise supplements exist.

And have a quick guess where spermidine got its name from? Yes, you guessed right, it was first identified inside sperm, but do not worry because it is just a small molecule that is found across nature, it is just bad luck where it was first isolated from (and then named).

Ref: Spermidine: a physiological autophagy inducer acting as an anti-aging vitamin in humans?. Autophagy. 2019;15(1):165-168.
Ref : Spermidine delays aging in humans. Aging (Albany NY). 2018;10(8):2209-2211
Ref : Novel aspects of age-protection by spermidine supplementation are associated with preserved telomere length. Geroscience. 2021 Apr;43(2):673-690
Ref: The autophagy enhancer spermidine reverses arterial aging. Mech Ageing Dev. 2013 Jul-Aug;134(7-8):314-20

Sirtuins and Sirt6

How does a 27% increase in lifespan sound? Well, the Sirtuin proteins are proteins responsible for keeping our cells young, and there are 7 known Sirtuin genes / proteins (genes make proteins). Sirtuin number 6, or Sirt6 seems to do much of the heavy lifting, and when mice were given compounds known to promote Sirt6 activity, those mice lived 27% longer than their normal counterparts.

- Can mice be used as an accurate guide for humans? Well yes, kind of, because out of the thousands of studies performed on those poor mice, only a handful performed on mice did not convey over to humans. Note I am not talking about drug research and toxicity trials, those are different, we are talking about gene activity here. So mice can be a good indicator for gene and protein behavior in humans (but only ever as a guide).

So, when some mice were genetically modified to overexpress Sirt6, researchers found that there was a 27% and 15% in males and females respectively (sorry ladies). But when overexpression of Sirt1 was induced at the same time, there was no further extension in lifespan, meaning that Sirt6 appeared to do all the work.

The mice with elevated Sirt6 levels appeared to have better cellular respiration, meaning they exhibited better metabolic plasticity (more food was converted into energy as opposed to being stored as fat), meaning the mice had more energy and were much more active in their later years, which could be interpreted that they remained more youthful (or at least experienced age delay). This higher cellular energy also protected those mice from much of the age-related decline that we see in older generations, because the cells could function youthfully. Interestingly, NAD did not decline in those Sirt6 older mice, meaning those old mice could manufacture NAD as they once did when they were young. This is huge news.

"Hurry up Ray, how do we activate Sirt6 please!!!". Well, Fisetin which is found in strawberries works, quercetin found in onions, broccoli, peppers, shallots, also Kaempferol found in spinach, kale, herbs, chives, fucoidan (from seaweeds), oleic acid (olive oil) and cyanidin (berries, black currants, grapes, etc). So far cyanidin appears to be the most potent of activator for Sirt6. Many of those activators can be found in supplement form, so rather than eat a mountain of those foods, supplement as well to ensure you are getting the elevated levels required for Sirt6. There is not much data on how much cyanidin you should take, but it is well tolerated and does not appear to cause any adverse reaction, however oleic acid is also a potent activator.

Oleic acid is hard to find as a supplement, but it is also found in olive oil, and is probably why the Mediterranean diet has offered longer lifespans for people, because it had a secret Sirt6 activator hidden within it.

Note that Sirt1 and Sirt6 are both upregulated during calorie restriction, and I have named one of the many studies below should you wish to go deeper on this topic.

A Comprehensive Analysis into the Therapeutic Application of Natural Products as SIRT6 Modulators in Alzheimer's Disease, Aging, Cancer, Inflammation, and Diabetes. Int J Mol Sci. 2021;22(8):4180.

Collagen

Anti-Aging Score 8/10

Collagen is a structural protein in skin and bone that makes up 25% of all the body's proteins. Collagen helps keep our tissues together, such as bone, teeth, tendon, ligaments, vascular ligature, and organs. There are 28 types of collagens, with collagen type I found in the above tissues, whereas collagen type II is found in cartilage.

So far, I cannot see that Collagen can be absorbed directly into the skin from the outside, and this may infer that skin creams that have fancy ingredients like collagen, elastin and more could be a total waste of money. However not all skin creams may be a waste of money, it just depends on how easily the active ingredients can be absorbed. Skin aging is accelerated by things like sunlight, smoking, environmental factors such as pollution, excessive alcohol, malnutrition and more. These facets of skin deterioration cause the skin to lose elasticity which accentuate fine lines and deliver more wrinkles. Loss of collagen causes the skin to become slimmer and drier.

A systematic review of collagen looked at numerous trials and found that collagen supplementation had beneficial effects on skin, but this was not done with standard collagen (which gets broken down inside the gut). They used a form of collagen called hydrolyzed collagen for improving skin health. The systematic review is not sure what the mechanisms were at work here (even though it sounds straight forward, I assure you the biochemistry is not), but either way hydrolyzed collagen has a beneficial effect on aged skin and damaged skin.

The review was called "Collagen supplementation for skin health: A mechanistic systematic review". Published 21st May 2020.

The hydrolyzed collagen is basically broken down into smaller molecules, and these smaller molecules are basically peptides. Peptides are interesting because they are known to act as signaling molecules, and therefore the mechanistic pathway of using hydrolyzed collagen is unknown; is the peptide telling the cells to make more collagen? Or does the peptide simply act as a substrate for faster collagen production? And there are a few other theories out there as well.

So how much should you take? The systematic review I have mentioned above showed that subjects were taking up to ten grams per day, which is a high dose, however you may not need that, or you may not need to take that amount for an extended period. So, if you have relatively good skin, 5 grams should be fine, but because I do not know what you look like, you will have to use common sense when deciding what dose and for how long you take that dose. Because hydrolyzed collagen comes in powder form, you could most likely add it to cereals (if you eat those), or other food groups, otherwise just drink it with a plant-based milk. Because hydrolyzed collagen is a tripeptide, I cannot recommend heating it up, as it may affect the stability of the protein, so do not cook with it.

Other beneficial effects from collagen is for osteoarthritis including the use of glucosamine, chondroitin sulfate, hyaluronic acid, vitamin D. This data was from a meta-analysis which is the highest standard of looking at multiple studies. What they found was that collagen was better than glucosamine in relieving symptoms of osteoarthritis. The study found a major reduction in the osteoarthritis index, and a significant reduction in stiffness. That study was called "Effect of collagen supplementation on osteoarthritis symptoms: a meta-analysis of randomized placebo-controlled trials" Published 27th October 2018.

And it is believed the way this works is that the tripeptides signal to the body to renew or regrow certain cartilage, ligaments, or tissues, strengthening and regenerating specific areas of the body. This could certainly be considered an anti-aging effect and could also be considered a method of keeping much of the body's tissues young (and in the event you do not even have any stiffness or osteoarthritis yet, supplement before you get that stiffness, prevention is always better). So, knowing all this about collagen it has become part of my anti-aging toolkit?

A final note about collagen, do you need to take 28 different types of collagens? No, of course not, just use the hydrolyzed collagen that appears to signal multiple tissues to regenerate. If you cannot find it locally, do an internet search on a well-known eCommerce platform, you will find what you need.

Astaxanthin

Anti-Aging Score 4/10

Astaxanthin is neither an amino acid nor an enzyme but a carotenoid that occurs in lobster, shrimp, salmon and other types of seafood which just happens to be a great anti-oxidant among other things. Astaxanthin is the pigment that gives those lifeforms their pink / orangey color.

Interestingly Astaxanthin is lipophilic which means it bonds to fats and oils and can also bio-accumulate. This is why the flesh of salmon and the exoskeleton's of lobster and shrimps are pinkish red. Astaxanthin is red in color and causes this effect as it builds up in the flesh and bone of living things, and yes, the science says that if you overload on Astaxanthin you could also start to have an orange or pinkish tinge which is known as beta-carotenosis.

But let's look at the benefits of this carotenoid.

- Resists oxidative damage on your cells
- 500 times more powerful than vitamin e
- 46 times better than lutein
- Prevents UVB-related sun damage
- 3 times better than beta-carotene
- Improves fertility in men
- Increases immune function
- Increases HDL cholesterol
- Protects against LDL cholesterol oxidation
- Prevents eye disease
- Reduces cancer risk
- Stabilizes blood sugar
- Easing symptoms of neurodegenerative disease

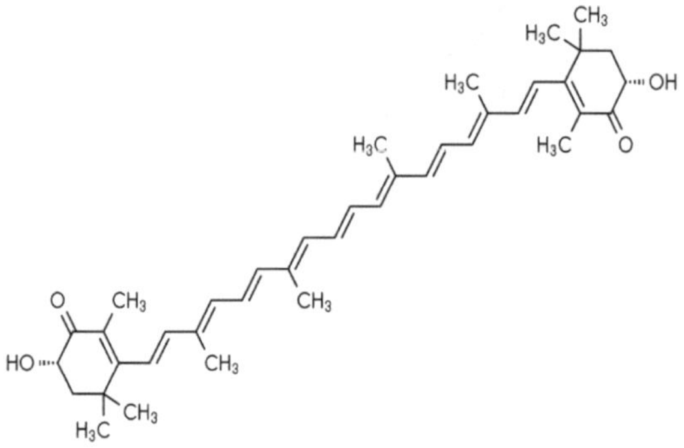

You should consume ~ 6mg Astaxanthin with a healthy fat or oil which helps it dissolve (more or less) and which then carries the Astaxanthin around the body (unless your version comes with an oil already).

We have already talked about vascular aging; Astaxanthin showed in a double blind placebo trial that Astaxanthin reduced oxidation of fatty acids which is believed to be the main culprit in heart disease.

Another placebo controlled randomized study showed that Astaxanthin reduced the buildup of something called "phospholipid hydroperoxides" which are considered a cause in dementia. Note that Astaxanthin also protects your mitochondria (which we will learn about a little later). However there is no magic pill, everything you read in this book to enhance your longevity will need to be in conjunction with other technologies and a healthy lifestyle.

GeroScience. 2017 Feb; 39(1): 19–32.
Mar Drugs. 2018 Apr; 16(4): 126
Mar Drugs. 2018 Aug; 16(8): 247

Deviations From Pharmaceuticals, Supplements and Enzymes

I am now and regularly going to deviate away from supplemental technologies such as those just discussed and throw in some information that at first may seem like a curveball. These types of deviations will happen throughout the book as you will see. However, these deviations are also connected to something else later to demonstrate certain points. So without further delay, let's keep empowering ourselves with knowledge and super charge that brain of yours.

Meditation

So deviating from longevity supplements, let's take a look at some interesting science about meditation. There is a growing body of evidence being drawn from very reputable scientific institutions that show the simple practice of meditation can have anti-aging benefits. But how could a state of mind increase life span? A study found at The National Library of Medicine demonstrates that genes responsible for inflammation exhibit decreased expression (resulting in less inflammation).

- The National Library of Medicine is where most published medical studies are located. It is an online database that anyone can access.

Note that the simple practice of meditation was also associated with faster cortisol recovery (which is important for your anti-aging toolkit as you will learn later in this book under the section Cortisol).

Can achieving a state of mind control how your genes communicate and control your body? It seems so. The team of scientists published their findings in the journal Psychoneuroendocrinology in 2013 which is referenced below. The team of scientists also acknowledges that the mechanisms at work here are still unknown and should be explored further.

Moreover, a study by Professor Kim Innes at West Virginia University found that meditation or a music listening program extended telomerase and reduced cognitive decline. The meditation group did better than the music group (however any increase in telomerase is a good thing).

State of Mind

So if meditation can affect our health, does this equate to our state of mind affecting our genome as well? The answer is absolutely yes. Your thoughts generate emotions and those emotions (according to science) affect the way genes express themselves. This is called epigenetics. So your mind is more powerful than you may think over your longevity. Maintain a peaceful non-stressful state of mind to ensure the genes for a long life are upregulated (turned on).

And it is worth mentioning one final study on this subject. Another team of scientists reviewed numerous other controlled studies that were performed on meditators and found similar results. The conclusion was that meditation could reduce inflammation, cell-mediated immunity, and biological aging (Black & Slavich 2016). This is quite game changing in our understanding of the mind body connection. Back to supplements -

Journal of Alzheimer's disease, JAD: October 2018
Kaliman, P. et al. Psychoneuroendocrinology. 2014 Feb; 40: 96–107.
Black, S & Slavich, M, Ann N Y Acad Sci. 2016 Jun;1373(1):13-24

Creatine

Anti-Aging Score 9.5!

This product is high on my list to prevent age related muscle loss. I consider age related muscle loss to be a serious hallmark of aging, and anything that prevents muscle loss is a "must use" product, especially for the older generation. We also have good amounts of human trial data, so we know it works, and works well.

Creatine is already found naturally in muscles, as it assists your muscles produce energy and lift heavy weights or perform high intensity exercise. Supplementing with creatine means that you increase your stores of something called phosphocreatine, which is mostly stored inside the muscles ready for use. Phosphocreatine helps create an energy molecule called ATP (more on ATP later), and this means that your muscles have more energy for explosive movements at any time. I am a daily user of creatine (5 grams per day), and yes, I can see a major difference in my training ability.

So when we think about aging, we usually think about people with impaired physical function, walking sticks, lack of strength, etc. Creatine helps in a few different ways –

1. Heightened Workload - Allowing more effort in a training session, meaning you can build more muscle preventing age related muscle decline.
2. Cellular Signaling – Improves communication between cells, so cells can repair and build muscles faster.
3. Cell Hydration – Creatine raises water content in your cells, which can play a role in muscle growth.
4. Protein Breakdown – Creatine appears to prevent muscle breakdown.
5. Inhibits Myostatin – Myostatin likes to break down muscles back to baseline, and creatine appears to inhibit this mechanism allowing for better growth.
6. Hormones – Creatine raise Insulin Growth Factor 1 (IGF-1) which is good for hypertrophy (muscle gain).
7. Brain Health – Creatine increases phosphocreatine in your brain, which promotes overall brain health and may ward off or slow neurological decline.

There are different types of creatine, and I personally use creatine-monohydrate, but most types all have the same effect, so don't over think it.

The paradigm we are entering is quite new here, as we now know that muscle loss and frailty in old age is partly (mostly) a choice, as you will see later. Lifting weights and maintaining strength is the magic bullet to maintaining youthful physical function and a better quality of life. Gone are the days when older people would sit all day knitting sweaters and watching daytime television. There are many people who are well into their eighties and nineties that have physiques and strength better than your average thirty-year-old, and you will read about this evidence further on in this book.

Am J Physiol. 1999 Sep;277(3):R698-704. doi: 10.1152/ajpregu.1999.277.3.R698
Proc Biol Sci. 2003 Oct 22;270(1529):2147-50. doi: 10.1098/rspb.2003.2492.
Exp Neurol. 1999 May;157(1):142-9. doi: 10.1006/exnr.1999.7049. PMID: 10222117.
Med Sci Sports Exerc. 2005 May;37(5):731-6. doi: 10.1249/01.mss.0000162690.39830.27. PMID: 15870625.
Int J Sport Nutr Exerc Metab. 2008 Aug;18(4):389-98. doi: 10.1123/ijsnem.18.4.389. PMID: 18708688.

Trimethylglycine TMG – Betaine

Anti-Aging Score 7
On the topic of muscle quality and strength, let's chat about TMG or Betaine.

Betaine is found in food such as broccoli, spinach, shellfish, grains and beets. It got the name betaine from first being discovered in sugar beets back in the 19th century.

Betaine is believed to promote creatine levels, so this may assist further if you are also supplementing with creatine as well. Remember, you want to resist muscle loss through age at all costs, so this is important. Betaine appears to increase the speed of ATP recycling, and again, we'll talk more about ATP later in this book. The more ATP your muscles have, the longer your muscles can work before fatigue sets in. Betaine can only give you these health benefits if you are performing high intensity interval training or lifting weights, it won't really kick in if you are lounging around on the couch binge streaming your favorite shows.

Brain and Heart Health – There is data that supports the idea that lowering an amino acid in your blood called homocysteine can delay or prevent the onset of neurological diseases such as dementia or Alzheimer's disease. High levels of homocysteine are also implicated in heart disease. Betaine or TMG which stands for tri-methyl-glycine has three (tri) methyl groups, and what this means to you is that TMG / betaine can convert homocysteine into another amino acid called methionine, and this appears (according to the literature) to lower your chances of neurological decline and heart disease.

Note that red meat and dairy products can raise homocysteine and you will learn that anyone serious about longevity will either reduce or avoid red meats altogether. Note the foods I have mentioned at the top of this page may help you to balance out any homocysteine raising foods you may be eating, so keep them in mind when eating red meat or excessive dairy.

The average daily dose of betaine could be anywhere from 100mg to 5 grams. So, do you have a history of heart disease or dementia in your family? Do you eat vast amounts of red meat or dairy? Think about those questions and chat with your doctor to work out the best dosing regimen for you. I take one gram per day.

- And worth a quick mention that betaine or TMG is also a great epigenetic regulator which resists epigenetic drift, which you will learn about later in the section called "Epigenetic Diet – Epigenetic Drift.

Ref: Betaine supplementation decreases plasma homocysteine in healthy adult participants: a meta-analysis. J Chiropr Med. 2013 Mar;12(1):20-5
Ref: Betaine restores epigenetic control and supports neuronal mitochondria in the cuprizone mouse model of multiple sclerosis. Epigenetics. 2020 Aug;15(8):871-886

Pterostilbene

Anti-Aging Score 3/10

Pronounced 'tero still bene', this great antioxidant is thought to be much more potent than resveratrol in cleaning up free radicals. Pterostilbene is close to the chemical structure of resveratrol but is known to be more bioavailable.

Bioavailable – The ability of your body to absorb something.

For our purposes in this book, we are focused on anti-aging technologies and pterostilbene's power in this sector lies in its ability to prevent cellular oxidative stress which prevents healthy cell death and ensures healthy cell function which in turn prolongs your life.

We could cover all types of antioxidants, minerals and vitamins in this book, however this book is targeted at compounds that have profound anti-aging effects and pterostilbene is too powerful to leave off our list. It is the primary antioxidant compound found in blueberries and is shown in vitro and in vivo to modulate neurological disease, to also be an anti-inflammatory, resist vascular disease and be an anti-carcinogen which is great if you're exposed to cancer causing carcinogens or just want to be healthier.

In Vitro – Experiments performed in a lab
In Vivo – Experiments performed on humans

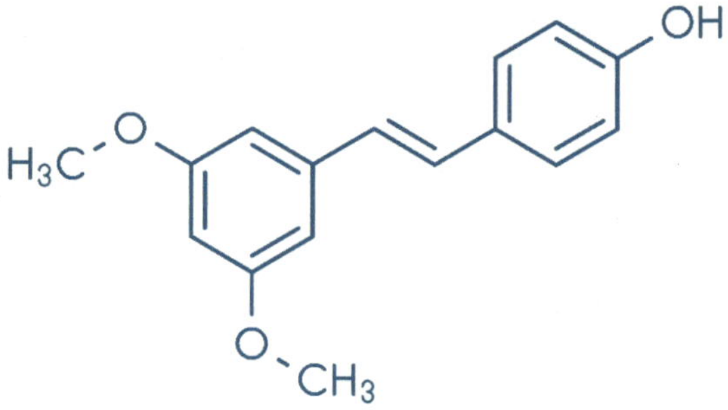

Animal studies demonstrated that pterostilbene was 80% more bioavailable when compared to resveratrol at just 20% (but do not discard resveratrol just yet as you'll read on the next page).

Pterostilbene has been shown to have an anti-proliferative effect on breast cancer in vitro and in vivo and also attenuates atherosclerosis (arterial plaque build-up). Pterostilbene protects against haemolysis (where red blood cells rupture) and protects against liver disease and continues in assisting with anti-cancer effects in cancers such as prostate, oesophageal, stomach, colon, liver and pancreatic. When we look at how powerful this antioxidant is, it is no wonder that I find pterostilbene a magnificent addition to my anti-aging repertoire.

But do not throw away your resveratrol so soon, anti-aging is a multipronged approach that requires numerous tools to control each aging pathway. This book is your tool kit to knowing which tools do what and how to use them. There is no point buying a hammer if you do not have the nails, and my recommendations here are no different; sure you can slow down the onset of one disease whilst you stay exposed to gaining another disease, but this book is aimed at providing you with a biological shield against aging across all known pathways, where possible.

Not all anti-aging supplements or foods are made equal. Many online options are of a low standard even though their manufacturers will state or claim that they are great. Using social media I will try and keep you abreast of which ones to use.

Pterostilbene has the ability to upregulate or downregulate pathways depending on the disease. For example it can promote cell death in cancer (apoptosis) or it can slow down apoptosis in cardiovascular disease (slowing or stopping the build-up of plaques).

Being objective as always in this book, pterostilbene may raise your LDL cholesterol (bad cholesterol), but in fairness to pterostilbene, your diet will be a major factor on whether it will raise your bad cholesterol. We will learn much more about diet later in the book.

McCormack and McFadden 2013,
Oxid Med Cell Longev, ID 575482
Evid Based Complement Alternat Med. 2014; 2014: 459165

Resveratrol
Anti-Aging Score Currently 1/10 More Data Needed

Resveratrol is a polyphenol stilbenoid that activates the Sirtuin genes. The Sirtuins are a group of genes that are believed to control facets of aging, with resveratrol activating the gene known as SIRT1. There have been some studies that appeared to show that resveratrol was all hype, however the research keeps coming in and the debate is still on. Resveratrol may be useful, and I say 'may', so I have included it here, but note, I do not currently rank it high on the list of molecules to use.

It is well known that resveratrol is found in red wine (from the skin of grapes), however you would need about 200 glasses of wine to get enough resveratrol to have an anti-aging effect, at which point you would be dead from the alcohol. So with your excuse to drink red wine utterly destroyed let's take a look at what we know so far.

Researchers at the University of Brighton and the University of Exeter (both of the United Kingdom) have recently found a way to reinvigorate senescent cells. Senescent cells are old dysfunctional cells which can no longer perform their tasks and increase your risk of disease. Notably in this study these cells had lost their ability to divide, however the cells also acted younger once subjected to the following analogue (a version of resveratrol). The researchers found that specific type of genes that switch off as we age and prevent further cellular division to repair and regrow the body could now be switched back on (and made youthful again).

This was achieved in the lab, but this data is game changing if it translates to humans. So what was the compound? It was a resveratrol analogue. And it did one more thing in addition to reactivating cellular division and restoring a youthful cellular state, it increased the telomere length, ensuring that cells could continue to divide (apparently infinitely if the resveratrol analogue were continuously administered).

- An resveratrol analogue is simply a comparable molecule to the resveratrol molecule.

Elderly people we see hobbling around frail and delicate are that way because their cells cannot divide further to regrow muscle, bone, brain, arteries, etc. Curing this state could be extremely close now.

Note that these results were achieved within hours of exposing the cells to the resveratrol analogues (not days, weeks or months) and the reaction was profound.

Professor Lorna Harries, Professor of Molecular Genetics at the University of Exeter, alongside Professor Richard Faragher and Dr Elizabeth Ostler from the University of Brighton are some of the bright sparks behind this discovery. The jury is still out on resveratrol though, as we lack human trial data.

Important - the study uses resveratrol analogues (resveralogues) and the resveratrol you purchase online may not be the same, so ensure you are getting the right stuff (which I will assist with later in this book).

If you'd like to learn more, the study was titled "*Small molecule modulation of splicing factor expression is associated with rescue from cellular senescence*" published in the Journal BMC Cell Biology.

Scientific studies can be a foreign language for the uninitiated, so beware, but persevere.

The Thermo-Neutral Zone

There is now science that indicates that being cold can promote longevity.

There are many animals that appear to live longer when exposed to lower temperatures. Kristin Gribble from the Marin Biological Laboratory (Massachusetts) has shown that even though exposing model organisms (Rotifers) to colder temperature there was a mean increase in life span from 6% to 100%, the results also concluded that genes played a big part in how much the cold would extend life span.

A particular theory on aging is known as The Free-Radical Theory on Aging. Free radicals are caused by metabolic activity or rogue electrons. The cold temperature theory predicts that colder temperatures reduce metabolic activity and therefore reduce the number of free radicals that go on to cause cellular decay and death.

However, I cannot recommend you subject yourself to cold temperatures, I am not your medical practitioner and everyone will deal with cold temperatures differently. If you are not sure how to keep yourself from freezing to death, then you should get professional advice first. You don't have to be super cold though; do not walk down the road whilst it's minus thirty in your jocks. You just need to be cold enough for your skin to react. Even a cold shower can have an effect.

Alternatively, the opposite may be true. Saunas and hot tubs seem to activate a protein (heat shock proteins) which can slow down or lower sugar spikes, and therefore keep your body closer to a fasting state (which you will learn more about later).

Ubiquinol aka CoQ10 or Co Enzyme 10
Anti-Aging Score 6/10

Often you will see supplements labelled CoQ10, however this enzyme comes in two types, ubiquinone and ubiquinol. Coenzyme 10 or CoQ10 is found in all cells of our body. In fact when we suffer from extremely low levels of CoQ10 it can cause a myriad of problems for our mitochondria and therefore our cellular function, at which point serious disease can creep in.

Ubiquinone assists with energy production whereas ubiquinol functions as an anti-oxidant that protects cells from free radicals.

Ubiquinol is more bioavailable than its brother ubiquinone, however both enzymes are worth implementing into your anti-aging regime.

Ubiquinol appears to be suppressed among older people with liver and cardiovascular disease therefore supplementing with these enzymes may slow or assist with these types of conditions. Ubiquinol has an important role in your cell's energy production by synthesizing a chemical known as adenosine triphosphate or ATP. ATP allows your cells to function at maximum capacity which ensures a lower chance of cellular mutations and disease.

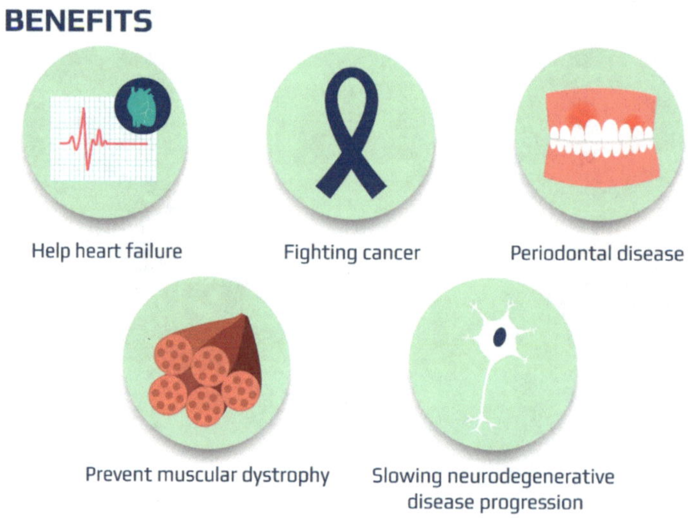

A study by Reiter et al 2009 shows that Coenzyme Q10 when used with Quercetin "reduced DNA damage after oxidative stress" and that "no cytotoxicity was measured". Keep this in mind for cancers that are a result of oxidative DNA damage. Do an online search of the study if you would like to learn more.

- Oxidative Stress is (basically) when free radicals are being produced inside the cell faster than a cell can detoxify itself (with antioxidants or neutralization) from those free radicals.

- Cytotoxicity simply means something that is harmful or toxic to cells.

Quercetin

Anti-Aging Score 6/10

Quercetin is the most ubiquitous bioflavonoid on the planet and the most consumed bioflavonoid by humans. So what is a bioflavonoid? It is the coloring or pigments that you find in many fruits and vegetables, so you could safely say that quercetin is a plant pigment. However quercetin has a powerful ability to scavenge free radicals from the body. This effect reduces DNA mutations that can cause cancer and reduces damage to cell membranes and cellular death.

Before you use it though, know this. Quercetin should be used synergistically with other antioxidants and bioflavonoids and not on its own. Genistein, apigenin, resveratrol, CoQ10 and green tea are good candidates. Some science shows that without using it synergistically that you could be wasting your time.

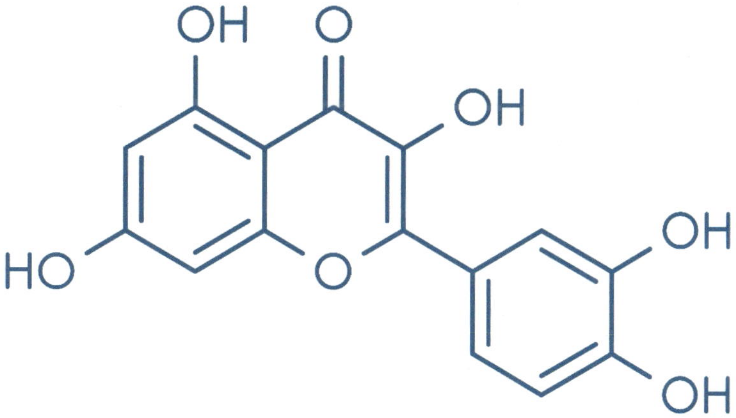

quercetin

Quercetin is an anti-inflammatory and works by decreasing pro-inflammatory cytokines which are shown to be a leading cause of premature aging.

Try and get the form of quercetin known as quercetin dihydrate for maximum bioavailability.

Ref : William, Spencer and Rice 2004, Free Rad Biol Med, 1:36(7):838-49
Ref : Aviram & Fuhrman 2002, Ann NY Acad Sci, 957:146-61

Sleep
Anti-Aging Score 10/10

Sleep is paramount to restorative and regenerative functions across the entire human body. Your body goes into repair mode the minute you fall asleep, and it is overwhelmingly documented in the scientific community that if you are not getting good quality sleep, you could be accelerating organ dysfunction, neurocognitive impairments and be at increased risk of mortality.

Sleep when your body tells you to. If you get the urge for a daytime powernap, take it, as the body is telling you it wants to perform a function. Seven to eight hours of sleep appears to be accepted amongst scientists, but if you need more then take it, but it could be a sign that something is wrong if you constantly require more sleep. Some body builders sleep twice a day to maximize the time their bodies can recover and grow.

The body must drop about 2 degrees Fahrenheit for sleep to take place, and this is why a colder room, around 65F/18C is better for sleep as opposed to a hot summery night.

Some random sleep facts worth knowing –

1. Men who sleep around five hours a night have smaller testicles.
2. Men who sleep four to five hours per night have testosterone levels equivalent to someone ten years older (you have aged a decade with long term sleep loss)
3. Female reproductive health is greatly affected by sleep loss.
4. After learning, sleep is like hitting the save button on a PC in order not to forget.
5. Sleep may get worse during aging.
6. Using sleeping pills does not offer the same benefits as natural sleep.
7. One hour of sleep loss (during the first day of daylight saving) sees a 24% increase in heart attacks. Conversely, when the extra hour is put back on, there is a 21% reduction in heart attacks. Similar trends with car crashes and suicide rates are noted here as well.
8. Those that sleep four hours on a single night, have a reduction in natural killer immune cells by seventy percent. This increases your risk of disease including cancers.
9. The World Health Organization has classified nighttime shift work as a probable carcinogen.
10. Sleep loss causes changes in gene expression, increasing inflammatory markers and heart disease.
11. Poor or shorter sleep increases all-cause mortality.
12. The shorter your sleep, the shorter your life.

Get some shut eye people!

Ref: Asian J Androl. 2014 Mar-Apr;16(2):262-5. doi: 10.4103/1008-682X
Ref: J Clin Sleep Med. 2018 Oct 15;14(10):1757-1764. doi: 10.5664/jcsm
Ref: Pflugers Arch. 2012 Jan;463(1):121-37. doi: 10.1007/s00424-011-1044-0
Ref: Medicina (B Aires). 2019;79 Suppl 3:29-32. Spanish. PMID: 31603840.
Ref: J Circadian Rhythms. 2020 Mar 23;18:1. doi: 10.5334/jcr.190.
Ref: Sleep. 2010 May;33(5):585-92. doi: 10.1093/sleep/33.5.585

Pyrroloquinoline Quinone aka PQQ

Anti-Aging Score 4/10

A coenzyme is a substance that heightens the operation of an enzyme. And an enzyme is simply a protein that acts as a catalyst to induce and or speed up a chemical reaction inside a cell.

Pyrroloquinoline quinone or PQQ for short is a coenzyme that incites mitochondrial biogenesis (the formation of new mitochondria in old cells). PQQ is another great product that should become part of your anti-aging toolkit. Other enzymes we cover in this book also target mitochondrial health; however PQQ creates entirely new mitochondria which is another game changing tool. You could look at this as like putting more batteries inside your remote control car, it will go faster and for longer.

However PQQ is also efficient and potent at removing hydroxyl and superoxide radicals (two of the most deleterious free radicals our bodies produce). To give you an example of how powerful some of the enzymes are in this book, PQQ is estimated to be 30 to 5000 times more effective than antioxidants such as vitamin C (ascorbic acid) in sustaining mitochondrial energy production (redox cycling).

The human body does not appear to have the ability to manufacture PQQ so you need to get it from diet or supplementation. 20mg per day has been shown to improve neurological function in people over 50, and the results became even more impressive at 300mg per day (however I do not recommend 300mg per day unless you are speaking with your Doctor first). PQQ also appears to show an extraordinary ability to minimize damage to the heart muscle during heart attack.

PQQ shows incredible neuroprotective and cardio-protective properties on these two all-important organs, but why? PQQ functions as a double pronged anti-aging weapon; it acts as a cell signaling modulator allowing cells to communicate with each other to ensure maximum biological function and at the same time it also acts as a superior anti-oxidant, as we have already shown above. Ultimately it is a heavy weight protector against neurological and cardiovascular decline.

To cover just some of the latest studies (which I will reference below), PQQ has been shown to prevent self-oxidation in the Parkinson's Disease gene DJ-1, which in turn protects against the onset of this degenerative disease. PQQ also seems to protect neurons and nerves from reactive nitrogen species (RNS). RNS can cause extreme stress on already damaged neurons from stroke or injury, however PQQ has been shown to suppress further damage from RNS and even provides further protection by impeding gene expression of another tongue twister known as inducible nitric oxide synthase (iNOS) which is also a major propagator of RNS. PQQ not only stimulates neurogenesis (nerve production) but also protects against toxins, especially mercury (so pay attention you seafood loving pescatarians).

Ref: Murase K et al 1993, Biochem Mol Biol Int. Jul; 30(4):615-21.
Ref: Hara H, et al, 2007, Neurochem Res. Mar; 32(3):489-95.
Ref: Zhang Y, et al, 2002, Neurosci, Sep; 16(6):1015-24.
Ref: Nunome K, 2008, Biol Pharm Bull. Jul; 31(7):1321-6.

Cordyceps - Invasion of the body snatchers?

Anti-Aging Score 2/10

This incredible fungus has the power to take over the minds of the insects that it infects. Once it infects its host it makes the host climb to the top of a tree, bite down with its mandibles, grows out of its skull and then release spores to the wind to infect other unsuspecting insects on the forest floor.

Fungi are one of the most potent life forms on the planet, so do not underestimate them. Keep your fridge door closed for three weeks and under total darkness with near freezing conditions, your fridge will thrive with fungi. After the extinction event that wiped out the dinosaurs the cold nuclear winter that plunged Earth into darkness for many years saw fungi become the dominant species on our cold dark planet. Fungi is more potent than you can imagine. So how do these fungi help you prolong your health or life span?

Cordyceps Sinensis (growing from the head of caterpillars)

Well there are over 1000 types of Cordycep fungi and here in this book you are only interested in Cordyceps Sinensis and Cordyceps Militaris. Cordyceps Sinensis can increase endurance in humans and assist in recovery by scavenging free radicals (Rossi et al 2014). That's important for those that are exposed to high levels of free radicals.

You might think foods such as coffee give you energy, but all they do is squeeze adrenaline out of your adrenal glands for a temporary kick. Cordyceps Sinensis upregulates the energy source (adenosine tri-phosphate or ATP) in your cells for a much more sophisticated fuel supply which ensures your cells function properly, keeping you young and disease free. It's also known to improve the symptoms of asthma (Wang et al 2016)

Fast becoming another fungal powerhouse is Cordyceps Militaris. A study from Hwan Hee Lee et al published in The Journal of Pharmaceutical Sciences showed significant delay in colorectal carcinoma tumor growth when exposed to Cordyceps Militaris.

An extract from Cordyceps Militaris known as Cordycepin also showed "very potent anti-cancer, anti-oxidant and anti-inflammatory activities" (Tuli, H. et al 2014).

Cordyceps Militaris

There are studies which show some leading Cordyceps products do absolutely nothing as well, so *caveat emptor* (buyer beware).

Lucky for you that Cordyceps cannot take over your mind just yet; they are completely safe, but keep an eye on your loved ones if they start supplementing and begin climbing trees.

Ref: Hwan, H, L. et al. 2015. DARU, 23, 35
Ref: Rossi, P. et al 2014. Evid Bas Comp Med ID 979613, 11 pages
Ref: Tuli, H, S. & Sandhu, S, S. & Sharma, A, K. 2014, Biotech, 4, 1-12
Ref: Wang, N. et al 2016. Evid Bas Comp Med ID 6134593

Melatonin

Aging-Aging Score 8.5/10

Many of you may already supplement with melatonin because it is a great antioxidant but I want to show you the correct way to use it and demonstrate the link and science behind how it works for longevity.

By now you are probably starting to see the word mitochondria quite a bit throughout this book. You can regard mitochondria as the batteries inside your cells and adenosine triphosphate or ATP as the chemical that powers those batteries. The more mitochondria you have in your cells the better, the more ATP inside your mitochondria the better. Mitochondrial decline is related to most (if not all) age related diseases, including obesity, so keeping those little battery packs working in their optimal state is extremely important.

Melatonin holds a unique power to fight mitochondrial dysfunction. Scientists found that a cause of mitochondrial dysfunction is a hole that appears on the inner membrane of the mitochondria which decreases their capacity to generate energy. Until very recently there were no wonder drugs (prescription or over the counter) available that had been known to plug this little hole, but a landmark study published in the Journal of Biochemistry and Biophysics in 2017 shows melatonin assists with maintaining the levels of an enzyme which plugs this hole. Worth noting that the hole is actually known as the "mitochondrial permeability transition pore". The enzyme (in case you want to learn more later) is known as CNPase.

The above study alone is reason enough to supplement with melatonin, but there is more (of course). Note that melatonin is a sleep hormone produced by the pineal gland inside the brain. It is comprised of an amino acid known as tryptophan and serotonin. So supplement at night time otherwise your circadian rhythm (sleep patterns) could be affected (which can open you to other problems). Just before sleep is the only correct way to use it.

Cancer - Melatonin has been shown to be a great anticancer supplement. I'll cover just a few facets of this field. Females that had metastatic breast cancer and had stopped responding to tamoxifen showed an improved response to chemotherapy when given 20mg of melatonin in the evenings. Prostate cancer cells showed a profound slowdown in replication when given melatonin; note that this was only in the lab, though promising results none the less. It was also shown that across all types of cancer that melatonin reduced the risk of death by a mammoth 34% (Mills, et al, 2005).

Migraines are also a victim of melatonin. A study shows that when supplementing with 3mg's of melatonin half an hour before bedtime, sufferers showed a 50% drop in the number of migraines per month. Not bad considering most expensive migraine drugs from the pharmacy cannot perform as well as this naturally occurring hormone (see Peres et al for more).

There is much more I could write about melatonin but this is simply a brief overview for your anti-aging tool kit.

Ref: Peres, M, F. et al. Neurology. 2004 Aug 24;63(4):757
Ref : Sainz, R, M. et al, 2005 Prostate Apr 1;63(1):29-43.
Ref : Mills E. et al, Journal Pineal Res. 2005 Nov;39(4):360-6.
Ref : Baburina, Y. et al. Biochim & Biophys. 2017. 1859(1):94-103

Curcumin and DNA Methylation

Aging-Aging Score 8.5/10

Curcumin is derived from the turmeric root and is the compound that gives curries their bright yellow color (do not confuse curcumin with cumin as they are different). Scientists have been intrigued as to why some countries that use a lot of curcumin in their cooking seem to have lower rates of cancer, with India being the prime candidate.

DNA Methylation is a mechanism where genes express themselves (basically genes being able to turn on or off) which instructs cells on how to act and function.

Scientists have found that most, if not all cases of cancer have a pre-existing condition, and that is that tumor suppressing genes are turned off. In scientific terms, if methylation occurs inside the promoter regions of a suppressor gene, tumors (mutations) can be induced. However lucky for us the opposite is also true, methylation can take place and ensure our tumor suppressing genes are turned on.

In a brief televised study three groups were chosen. The first group was given a placebo, the second group was given a turmeric pill and the third group was given turmeric powder and asked to cook with it. The placebo group showed no change in their gene expression and neither did the pill group, however the group that used curcumin powder in their cooking showed something incredible. DNA methylation had occurred and their tumor suppressor genes were turned on, as if providing the subjects with some type of genetic armour.

The science of why turmeric when pressed into a pill did not have the same effect is unknown, but it could be because the pill took too long to dissolve inside the stomach reducing the effect, or that cooking with turmeric made it more bioavailable for the third group. However the result on gene expression is known and curcumin should be in your anti-aging toolkit.

Curcumin is lipophilic (it binds to fats), so when cooking with turmeric the oil assists with absorption. Add black pepper (pipperine) to this meal to enable better absorption. In the very near future you will be able to take a test to see if your suppressor genes are on or off and make dietary changes to control them. Food is medicine and food is poison so choose your weapons carefully (most processed and animal based products will speed up mortality). Being able to predict cancer up to two years in advance is one of the biggest things in medical history, but you've probably never heard of DNA methylation until now. You can thank The University College of London (UCL) for this ground-breaking science.

P.S I use a table spoon of coconut oil and a gram of curcumin powder, make a paste, and then add that to my coffee every morning without fail.

Ref : Beck, S. & Rakyan, V,K. 2008, Trends Genet, 24, 231-237
Ref: Mosley, M. et al, BBC. Trust Me I'm a Doctor, S5, E3
Ref: Reuter, S. et al, Genes & Nutri, 6, 93-108

Berberine

Anti-Aging Score 8/10

Dubbed by some as the next Turmeric (Curcumin), berberine has a range of benefits worth including into your anti-aging toolkit. I could write a few pages on this phytonutrient (a compound produced by plants), but I'll keep things quick and direct.

Berberine can assist in regulating the correct amount of calcium into the cells, as calcium acts as an electrolyte which controls the electrical signals that control the heart, which equates to berberine functioning as an anti-arrhythmia compound. This in turn can improve heart performance and at the same time decrease blood pressure, heart attack risk, and decrease ischemic episodes (the blocking of oxygen to tissues) due to arterial constriction or arterial plaque buildup. And note research shows berberine assists in reducing arterial plaques (atherosclerosis) and when co-supplemented with fenugreek seeds, assisted in lowering bad cholesterol (LDL). Berberine can also act as an anti-coagulant which assists in preventing clotting (possible stokes) by keeping the blood at the right viscosity. Berberine can also reduce the effects of caffeine, which are a cause of heart arrythmias. So that's some serious "heart health weaponry" for your anti-aging toolkit.

Berberine also appears to act as a naturally occurring antibiotic and can assist in immune function and has been shown to have anti-tumor and anti-cancer properties. Data suggests that berberine can effectively lower blood sugars which can help with insulin resistance. On this note, it is worth saying that there is serious research now going on in relation to whether berberine is better than metformin.

And of course, berberine is a potent anti-inflammatory, and this is where it may compete with turmeric / curcumin. Inflammatory conditions such as bursitis, tendonitis, joint inflammation such as arthritis all seem to respond to berberine well, so it's worth considering what your family history is, your current health status and whether berberine could add value to your biology.

I take 500mg per day primarily for inflammation and atherosclerosis (based on family history), but if you have atherosclerosis, then data does indicate that taking two 500mg doses per day (one in the morning and one in the evening) works better.

Ref: Gaita F. Berberine behind the thriller of marked symptomatic bradycardia. World J Cardiol. 2013 Jul 26;5(7):261-4
Ref: Min wu, Front. Pharmacol., 09 July 2021 Sec. Inflammation Pharmacology Volume 12 – 2021
Ref: Anti-inflammatory activities of berberine in the treatment of diabetes mellitus. Evid Based Med. 2014;2014

Adenosine Monophosphate-Activated Protein Kinase aka AMPK
Anti-Aging Score 9/10

I will show you a few more products worth supplementing with and then we can move onto diet, exercise and some other game changing technologies worth a mention.

AMPK is a cellular enzyme that if lacking is arguably involved in most if not all disease. The lower your AMPK activity, the more chance you have of developing disease. Information is power, and knowing this particular science means that you can now greatly affect your chances of developing disease as our biological clock sleep walks into old age at one second per second.

Professor Emeritus William Clark of Immunology at UCLA says "AMPK is a suppressor of aging itself". That's how powerful some in the scientific community regard the effect of this little enzyme and its control over aging. Scientists now have a huge body of evidence showing that this enzyme is one of the final powerhouses that anyone serious about longevity should be including in their anti-aging toolkit.

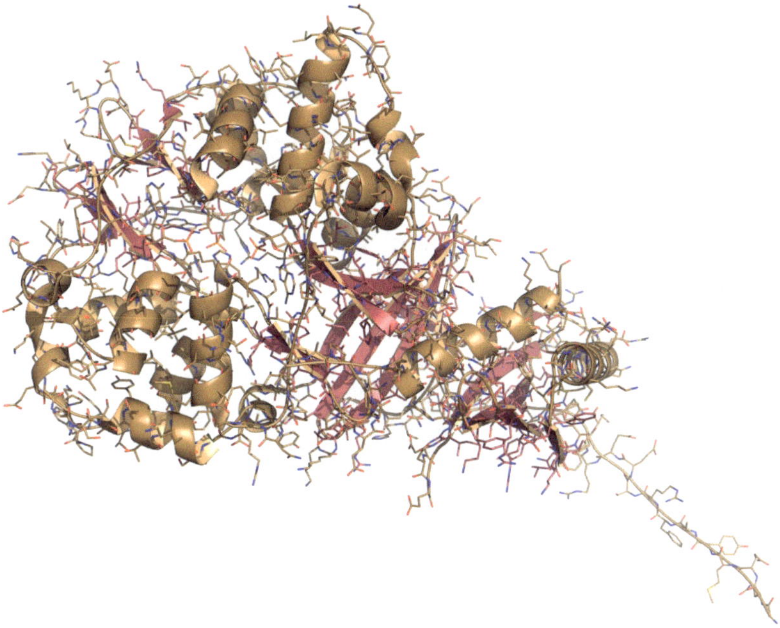

AMPK Enzyme Bound with AMP

Let's take a quick look at how it works. AMPK can sense the energy level of each cell (trillions of cells), which can then activate a response mechanism that tell the cells to get back into the most perfect energy state. Too little energy and the cell dies which can cause disease, too much energy can cause improper cellular function which can also induce disease. Think of cellular dysfunction as disease. All those elderly people you see with age related disease have little more than the result of cellular dysfunction.

As we eat food, energy is passed down a stream of enzymes until it reaches the battery pack inside our cells known as ATP (which we have covered earlier). As the ATP is used to power the cell, the end product (the ATP exhaust fumes) is adenosine monophosphate (AMP). As the cell fills up with AMP and depletes its supply of ATP, the cell becomes low on energy, possibly unable to perform functions or worse. Rising levels of AMP inside the cell biochemically activate the release of AMPK, and this is why AMPK is so important.

When new AMPK is released, it escalates 'fatty acid oxidation' (which is simply a process that creates new ATP). AMPK also initiates glucose transport, where sugars are taken from stored fats as additional energy sources for the cells to function. The net result is perfect cellular metabolism and flawless cellular machinery.

Fruit flies genetically modified to have more AMP in their cells live up to one third longer than their natural counterparts due to the induction of AMPK. People that eat junk food and live sedentary lifestyles will lower their levels of AMPK, inducing more fat storage, slowing cellular metabolism with an increased chance of disease.

The science is profound and the jury is in, AMPK levels are essential for your anti-aging toolkit in assisting you to avoid disease and maintain a youthful state.

And more good news, metformin and berberine both help activate AMPK (chat with your doctor about using them at the same time though, as they both control blood sugars).

Ref: Xu XJ, et al, Cell Cycle. 2011 Oct 15;10(20):3447-51
Ref: Stenesen D, Suh JM, Seo J, et al.. Cell Metab. 2013 Jan 8;17(1):101-12

Alpha Ketoglutarate AKG

Anti-Aging Score 6/10

Alpha Ketoglutarate (AKG) is used inside the cell to manufacture amino acids, create energy and rewind your epigenetic clock (more on epigenetic clocks later). Like many other biomolecules that keep us young, AKG tapers off with age, causing an energy loss inside the cell, which in turn causes the conditions we see with old age.

- I should note here that my research clearly shows me that every condition in old age can be traced back to some type of cell dysfunction, which much of the time can be attributed to lack of cellular energy. Keeping cells youthful (in my professional opinion) prevents many facets of aging. In fact, everything we talk about in this book will come back to some type of cellular restoration or replacement, rejuvenation, or removal of either dysfunctional cells or cellular junk (cells are the key).

And replacing AKG is easy with basic supplementation. However, do not go out and purchase it just yet. AKG has been shown to extend lifespan in rodents, flies and mice and prevent obesity in mice (probably from allowing cells to metabolize more food for energy), it also creates brown fat (the healthy fat). The first reference below also captures some great data for AKG, but note, even though that paper shows modest life-span extension (about 10.5% on average), what it did capture was that the duration the mice stayed fit and healthy, was longer than the time it extended their lives (about 40%). This means that it kept them younger for longer, they did not enter an old age state, and lived much healthier and vibrant lives, whilst their rodent counterparts dwindled and died earlier as nature intended. However, this product was slightly different, it was Calcium Alpha Ketoglutarate or CaAKG. And there are some idiosyncrasies between sexes; CaAKG did decrease inflammation in female mice, but not males, and this is because females produce a higher level of a cytokine called IL10. So note female readers, if you have aches and pains (inflammation), then CaAKG or AKG could assist you (sorry guys).

And note, AKG is not found in a normal diet, however it is boosted by a Ketogenic diet (more on Ketosis later) and can be boosted by supplementation. I take about two grams of AKG powder every morning.

The second reference below also captures that calorie restriction maintains AKG levels, again more evidence that eating more food than you need removes the longevity magic that happens deep within your cells.

Admittedly human trials need to be conducted, but the animal models are promising enough right now for me to include AKG into my toolkit. But which should you purchase? CaAKG or AKG? Well, the studies show promise with both, however AKG is what is used inside the mitochondria for energy (The Krebs Cycle), and not CaAKG. So, my knowledge tells me that AKG is a smaller molecule, and most likely easier to get into the cell and can be used immediately without the Ca having to be broken off. Note that biochemistry is a rabbit hole though, and the Ca may also assist in cellular entry, though I do not have side by side human trials to go off, so either should work, but I am comfortable using AKG until I see more science.

Alpha-Ketoglutarate, an Endogenous Metabolite, Extends Lifespan and Compresses Morbidity in Aging Mice. Cell Metab. 2020 Sep 1;32(3):447-456.e6.

Caloric restriction increases ketone bodies metabolism and preserves blood flow in aging brain. Neurobiol Aging. 2015 Jul;36(7):2296-2303.

Calorie Restriction / Fasting on Senescent Cells

Professor Valter Longo from the University of Southern California has been studying the effects of calorie restriction (CR) for many years, especially on yeast. Yeast cells respond to a low calorie diet by fighting toxins and protecting cells resulting in a life time of up to 10 times longer. How does that relate to humans? Research is showing that CR instigates potent natural healing processes that burn stored fat, decelerate aging, fight heart disease, lower your cholesterol and reboot immune system cells. But if that is not enough for you, fasting has also shown that it can increase the effects of chemotherapy and even be as effective as chemotherapy if we are to believe the data these trials are delivering (and I certainly do believe the data).

Humans appear to have extremely powerful self-healing mechanisms entombed deep down inside our DNA which are only ever actuated during periods of prolonged fasting. Those that eat regularly and enthusiastically their entire lives may never trigger those healing mechanisms and will rely on big pharma to do what calorie restriction may have done naturally. For example, even cancer cells can die as a result of calorie restriction because fasting restricts their supply of glycogen which they require more than normal cells for survival, thus those cancer cells are swindled into releasing damaging free radicals which causes those cancer cells to die prematurely (never cease conventional treatments for fasting unless instructed by your Doctor).

Still not convinced to change the way you eat? Professor Longo has shown that fasting induces the creation of new stem cells which can repair the body's tissues, repair organs and blood and kick-start the immune system.

Professor Longo has developed a diet which you can purchase in various regions. This diet will trick your body into thinking it is starving when in fact you are just hungry. During this five day diet your organs will shrink getting rid of any bad cells that are taking up precious food resources. When you start eating again your body renews those organs back to their original size, but this time with brand new stem cells, it's almost like our body is an embryo being built back up with new stems cells again.

This is an extremely scientific diet and sure you may be able to replicate the calorie intake, but the foods you choose will have to yield nutrients for that five day period. Ensure to take a multivitamin during any fasting period and do not binge on junk food once it is finished.

I will put a table on the next page for you to look at. This list of food is a four and a half day fast that I conducted. Note, the table does not show Professor Longo's diet, but a simple CR diet that I may try and do once every two months.

Doing the 5 day CR program once every two or three months would be adequate, there is no need to do this every week or every second week

Imai et al, Nature, 403, 795-800(17 February 2000)

A Typical Calorie Restriction Diet

The diet outlined below was conducted by myself for 4.5 days.

Day	Calories
Monday	
Black Coffee	5
Half avocado	70
Nut Bar	89
Sweet Potato Soup	118
Half avocado	70
Mint tea	2
20 grams walnuts	131
Vegetable Soup	84
9 Olives	70
Tomato Soup	111
Total Calories	750
Tuesday	
Mint Tea	2
Nut Bar	89
Mint Tea	2
Half avocado	70
Tomato Soup	111
Kale Chips 10 Grams	54
5 Olives	40
Spearmint tea	2
Half avocado	70
Pumpkin Soup	118
2 Olives	16
Spearmint tea	2
Vegetable Soup	84
5 grams almonds	32
Spearmint tea	2
	694
Wednesday	
Spearmint tea	2
Nut Bar	89
Half avocado	70
3 Olives	24
3 Crackers	81
5 grams almonds	32
Pumpkin Soup	118
Half avocado	70
1 cracker	25

Mint Tea	2
4 grams popcorn	15
Vegetable Soup	56
3 Crackers	23
Popcorn	30
Vegetable Soup	56
Total Calories	693

I spread my miniscule calorie intake out throughout an entire day. This helps shut down any cravings. However after a few practice runs, the body adapts and can handle this bi-monthly regime easily.

Thursday	
Spearmint tea	2
Nut Bar	89
Expresso	5
Half avocado	70
2 almonds	14
Vegetable Soup	56
2 Crackers	17
2 Olives	16
1 cracker	8
Popcorn	30
1 cracker	8
Vegetable Soup	56
1 cracker	8
Half avocado	70
1 cracker	8
Spearmint tea	2
3 walnuts	21
3 Olives	24
1 cracker	8
3 Olives	8
1 cracker	8
Vegetable Soup	56
1 cracker	8
Beef Soup	51
Cruskit (sandwich cracker)	42
Total Calories	685

Friday Until Lunchtime

Nut Bar	89
Half avocado	70
2 Crackers	17
Hibiscus Tea	4
Strawberry Tea	4
10 grams walnuts	65
5 Crackers	40
Total Calories	289

The total calorie count here for 4.5 days is 3113 calories. This caloric intake is less than what a standard Western diet may throw at you in a single day.

As you can see the food intake for each day is extremely small. Maintain this for five days and the effects on your cells will be profound. The tables are a screenshot from my own CR diet and are good examples to get you started on five day fast. You can vary the foods, but maintain the caloric limit. The more you practice CR the easier it becomes. Willpower is like a muscle, the more you use it, the stronger it becomes.

The 5.2 Diet

The 5.2 diet is for those of you that cannot do the entire five days of CR. So instead of five straight days of 700 calories or less, you can semi fast with 500 calories or less for two days out of seven. The 5.2 diet directs participants to have two days per week, maybe Monday and Thursday where they eat no more than 500 calories on those days.

Alternate days is another CR diet with benefits, however only the five days straight CR diet will induce new stem cells, the other two diets (5.2 and Alternate) are yet to demonstrate stem cell creation though they are proven to lower cholesterol, blood pressure and reduce fat.

As a species we developed when feast or famine was standard, we never ate three or four meals per day 100,000 years ago, so replicating that dietetic environment at a cellular level, where your system is being stressed is a good thing because our bodies are designed to be stressed in order to discard old cells and release new cells to build new tissues etc.

So what is the effect of this? Well new cells means better function, also means better disease fighting and more energy for you. After a few weeks of CR for either 5 days straight once a month (as per the section earlier "Calorie Restriction / Fasting and Senescent Cells") or for two days per week as per this section, you will feel better and abound with more energy due to your new born cells functioning as they should. You may not experience more energy the first time round, you will find it hard even, but with more and more cells being renewed and released into your system, you will begin to function at an optimal rate, the way nature intended.

During CR the brain starts to produce a peptide called brain derived neurotrophic factor (BDNF), which is basically a collection of proteins or peptides that induce neurogenesis (the creation of new neurons) which in turn have been shown to make you feel better and sharper. On this type of diet, if you eat normally on the other days, you should be able to shed 500 grams of fat each week. Many diets may show that you can lose kilos in the first week, but that is only water retention, not fat loss, though water retention loss is still a good thing as losing your water places less stress on the outer arterial walls lowering blood pressure (among other things).

Insulin Like Growth Factor or IGF-1 and Red Meat

The World Health Organization has included red and processed meats as carcinogenic.

IGF-1 or Insulin Growth Factor 1 is a hormone that speeds up cellular division, it makes your body grow, which is great if you want to be big and burly, but not so great if you want a long life. You want your cells to repair and stay around for as long as possible before dividing due to their divisional limits. However IGF-1 and even human growth hormone (HGH) which induces IGF-1 will cause you to plough through your cellular limits faster.

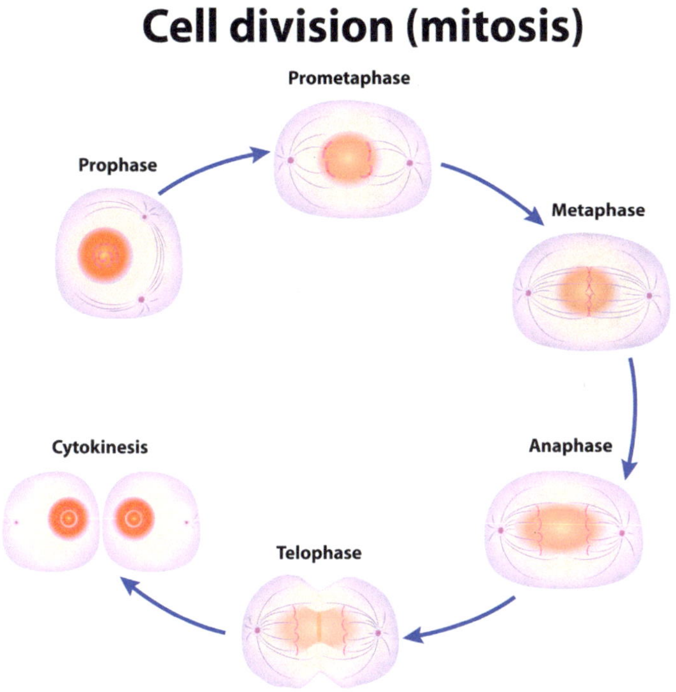

Animal protein elevates your levels of IGF-1 which accelerates the aging process and affects life span adversely (not to mention contains cancer causing PAH's and AGE's which we have already covered under the section "Sugars and Glycation").

- PAHs – Polycyclic Aromatic Hydrocarbons (the blackness on burnt meat)
- AGEs – Advanced Glycation End Products (keeping it simple, means premature aging)

Calorie restriction will lower your IGF-1 levels to slow down cellular division, but if you are eating a lot of animal protein during CR you could be wasting your time. However simply by going vegetarian / plant based you can induce similar effects of fasting or CR.

If you are releasing HGH from exercise, this is okay, because the exercise is lengthening your telomeres at the same time, but if you have been duped into thinking injecting HGH is the fountain of youth (as some claim), then you are completely wrong, because this type of HGH increases your IGF-1 without lengthening your telomeres, ploughing toward your cell's divisional limits.

The amount of times your cells can divide is called the 'Hayflick Limit' which you will learn more about later in this book.

The Effects From Red Meat IGF-1.

A great study published in the online journal Cell Metabolism profoundly demonstrates that a lower protein diet is associated with a major reduction of IGF-1, lower risk of cancer and overall lower mortality in people less than 65 years of age. The mortality reduced even further when the protein was changed to plant based protein.

Ref: Morgan, E, L. et al, 2014, Cell Metab. Mar 4; 19(3): 407–417

When I was growing up I remember being told many times that we required about 30 grams of protein per meal; and that is one of the biggest lies ever told. All you need to consume are foods that are rich in amino acids, which are the building blocks of protein. Gorillas, elephants, rhinos and other large land animals do not supplement their leafy green diet with protein shakes to grow that amount of muscle, so do not be tricked into an early grave by the food industry's clever marketing. Evidence shows that meat (for most people) may give you poorer health, disease and shorten your life. Eat some, preferably white meat, enjoy the benefits, but stay close to vegetarianism.

As a species we usually fast for around 12 hours per day already with our last meal being around 7pm and our first meal being around 7am. Snacking at night and reducing this fasting time is counterproductive for your anti-aging toolkit. During sleep (fasting mode) your body goes into repair mode whereby it rids the body of old dysfunctional cells; cells that can no longer assist with fighting disease and may actually cause disease, however the minute you begin eating again your body goes into the exact opposite mode. The longer you can go without food, the more repairs take place. Eating regularly and snacking in-between meals is slowly killing you through a complex network of biological pathways.

- Processed meats (deli meats) are much worse than normal meats as well. Most cancers and disease from meat, come from processed meats.

The Homer Simpson Effect - Gluttony & Fasting - Intelligence

Remember that CR = Calorie Restriction

Love it or hate it, the science speaks volumes. Scientists at the National Institute of Aging have found that putting mice on a CR diet where every second day they eat nothing actually increased brain function. The research was led by Professor Mark Mattson who is one of the world's leading authorities in this field. He found that these CR mice had around a 50% increase in brain derived neurotrophic factor (BDNF) which is thought to play a role in preserving nerve cells.

When we are hungry our stomach releases a hormone called ghrelin which binds to the hypothalamus. So in another study Dr Tamas Horvath of Yale Medical School injected mice with ghrelin and observed that these mice outperformed mice with normal ghrelin levels in cognitive abilities.

CR has been around since the year 1000 where a Persian polymath named Avicenna taught older people to consume less food than young people. In 1500 a Venetian Luigi Cornaro also practised CR and lived to 102. With CR now known to increase brain activity (making us smarter), it could be thought that nature has put this mechanism deep into our DNA, so that when foods are scarce, we become smarter and more innovative to derive ways to find food to ensure the survival of our species. Without this behaviour being programmed into our genes, who knows if humans would have survived when faced with times of famine.

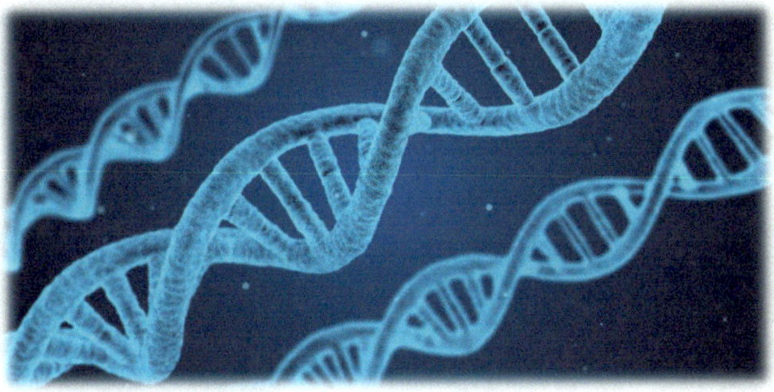

And the opposite seems true. Overeating can cause conditions where cognitive function and memory is impaired. This study was published in the Journal Neurology in 2006 by French scientists. The study took adults between the age of 32 to 62 and gave them four mental tests and then repeated those same tests 5 years later. People with of a body mass index of 20 or less (lean) could remember 56% of words from a test, whereas those that were obese could only remember 44%. The heavier subjects also showed an increased rate of cognitive decline, which is extremely worrying if you do not want to end your life with a neurological disease.

Another 25 year study found similar results and even showed that the early onset of obesity during childhood can have dramatic effects on IQ for life. However this does not mean overweight people are less intelligent.

Either way CR remains your best dietary option on all fronts for health, longevity and to increase intelligence.

Ref: Belsky, D. et al, Am J Epidemiol. 2013 Nov 1; 178(9): 1461–1468

Cortisol

Known as the aging hormone, cortisol has wreaked havoc on human health for millennia. Known to increase fat deposits, increase blood pressure, increase blood sugar levels, will cause hardening of the arteries, lower human growth hormone levels, accelerate muscle loss, may play a role in osteoporosis and is suspected of even causing a weaker immune system.

What causes cortisol? Basically it is a stress response hormone. In small doses it provides functions that help keep us alive, but prolonged stress over weeks and months can have a detrimental effect on our health. Those that experience long term stress with work, family, friends or even lifestyle choices may be walking themselves into an early grave not to mention having the signs of aging become more visible on their faces prematurely. This is why it is important that you only surround yourself in an environment with people that bring positive energy into your world. Negative people that always seem to drain others are also making you age faster. That's not to say cut off the mother in law and anyone else that you may not like, but certainly do not be around stressful people or in stressful environments on a regular basis unless its family or unless you are getting paid.

The type of long term inflammation that cortisol causes accelerates your aging in a process that people in anti-aging circles call inflammaging. Be aware of what causes your mind and body stress and have limited contact with it. Note that curcumin is a great anti-inflammatory. Also flax seed and leafy greens are also great anti-inflammatories.

Exercise

We all know that exercise promotes a healthy heart and has many other beneficial factors. What we are concerned with though is what type of exercise extends our life and what type of exercise could speed up aging. Earlier in this book we talked about telomeres and how important their length is on cellular aging. Well it turns out that exercise lengthens them.

A study carried out at Brigham Young University, Utah, United States by Professor Larry Tucker profoundly showed that individuals that performed regular vigorous physical activity had longer telomeres than those that sat down and lived sedentary lifestyles. But how profound were the telomere differences? For those that did intense exercise such as jogging for 30 or 40 minutes per day for 5 days per week had a 9 year advantage to those that did nothing. Those that did moderate exercise had a 7 year advantage. You do not have to jog for 30 minutes each day to live longer, just be more active around the house, in the garden, with the kids etc.

High Intensity Interval Training - HIIT

Better than jogging was HIIT. A study by the Mayo Clinic showed that HIIT did not just gave us all the physical advantages you would expect from exercise such as muscle mass and strength, improved cardiovascular function, decreased risk of diabetes, but our mitochondria (our cells battery pack) was boosted by 69%. When mitochondria slows, our cells cannot function properly leaving our body weak, frail and open to disease, and we've all seen older people hobbling around like this. Maybe had they done some HIIT they could have powered their cells to ward off any age related decline or disease. The Mayo Clinic's study showed that just two or three sessions per week can have dramatic effects on your mitochondrial function. So get that bike, get that skateboard, get that tennis racquet, get that ball and work those cells.

In a joint effort by the University of Mississippi and University of San Francisco, California where data was taken from participants and compared based on their exercise versus their telomere length. What was found was incredible. The more types of exercise that someone performed, the bigger and better the result. So for example, riding a bike is good, but lifting weights on the following day as opposed to just riding a bike was even better, then boxercise on the third day would be even better. Those exercises are just examples but you get the picture, so mix things up and do not use the same muscle groups over and over.

Worth noting as well that the age group that had the most benefits when performing different types of exercises were between 40 and 65, which means middle age could be the best time to get longevity savvy. Marathon running and similar exercise however can age you faster by increasing your cortisol levels for long periods, which also causes prolonged exposure to free radicals, in addition exposing you to the sun and increasing wear and tear on your joints and for many of these athletes eating more animal protein will also age you faster. So no five hour bike rides please, 30 to 45 minutes (at most) is fine and all your telomeres need.

- A fleeting thought - How lucky for you, I stand here at my standing desk (I never sit) for hours on end, researching and writing this book. Why? I just could not keep explaining all this data to everyone I met. Once people know you are a longevity researcher, they latch on with a million questions, now I just say "buy the book".

Lifting Weights

When it comes to lifting weights, I can almost guarantee you that if you haven't been trained properly to lift weights, there is a 99 percent chance you are lifting them wrong. Everything from warming up, to correct hand grip and movement, such as pronation or supination all get factored in to which part of the muscle is targeted. I have seen numerous people hit the gym for the first time only to be so unaware of how to lift, that they are on the "lets get an injury super highway".

Technique is everything when lifting weights. So again, either get a personal trainer or other trained professional that can teach you the basics. It is an investment in your health and longevity, and if you really care about your future biology, and by reading my book you probably do, a few lessons with a personal trainer will be an incredible and worthwhile investment for you.

Stretching – Building muscle is not enough during age either, you need to keep those muscles as flexible as possible, because as we age, we lose muscle range, and this equates to poor physical function and increases your chance of injury. As I hit my forties, I could no longer swing my arm from the driver's seat into the backseat to grab something without my shoulder going pop. I now stretch every joint and endeavour to claw back muscle range. It's really easy, takes a few seconds and pays big dividends, and you can do it whilst watching television or during a phone call etc. Whenever I have a few seconds standing around, I am also stretching something out. Make it a lifestyle, but chat with your doctor or a physical therapist on what is right for you (and there are many professionals on YouTube teaching you the basics on how to stretch properly).

EPOC - The After Burn Effect

Exercise Post Oxygen Consumption (EPOC) is basically a fancy way of saying that your body continues to burn calories after you have exercised. The more intense your exercise, the longer your body stays in EPOC, burning calories after you have exercised. This is why high intensity interval training on the previous page is so effective, because your body may continue to burn calories for many hours after you have stopped exercising. As opposed to an exercise such as walking, which would hold very little EPOC advantage, if burning / losing fat is your goal, then choose exercises that will drag out the after burn effect for many hours.

VO_2 Max

VO_2 max is your body's maximum oxygen consumption possible during physical action. V is for volume and O_2 is the term for oxygen (of course) and max is for maximum. When it comes to longevity, those with a lower VO_2 max die up to five times more than those with a higher VO_2 max. This basically means that fit people die five times less than lazier people.

This was demonstrated in a 2018 study which clearly showed that the less fit you were, the more likely you were to die earlier over a decade. This study split people into five groups from "elite" (in the top 5%) all the way to "low" (the bottom 25%). The elite had the lowest death rates, but the categories just below elite such as high, above average and below average still scored well and were actually still quite close to the elite level. This means that you only need to do moderate amounts of exercise to make yourself five times less likely to die. Beneath the below average category was the low category which showed an incredible increase in all-cause mortality, and you do not want to be in this category, so get on that bike, do sixty seconds of mountain climbers, do something so that you are out of breath a few times per week (your chances of dying will go down by a sizable margin).

Ref: W. Association of Cardiorespiratory Fitness With Long-term Mortality Among Adults Undergoing Exercise Treadmill Testing. JAMA Netw Open. 2018;1(6):e183605

Ref: Teixeira IP. Is the VO2max that we measure really maximal? Front Physiol. 2013 Aug 5;4:203.

Ketosis and Ketones (Ketogenic)

Something worth knowing is ketosis and how to manipulate it. Your body is a machine that requires fuel. If you are not feeding your body with carbohydrates, then your body needs fuel from elsewhere, and that will be your fat reserves. While this sounds good and can easily be achieved by eating plant based proteins without carbohydrates, if you are depriving yourself of carbohydrates and exercising energetically then your body may start to produce an excessive amount of ketones (more on ketones soon).

Ketosis (fat burning mode) does not kick in immediately after reducing carbohydrate consumption, you have to eat very little carbohydrates for a few days (possibly less than 500 grams of low sugar (low carb) food per day for three days) for the noticeable fat burning to take effect. Even though 500 grams of low carbohydrate food sounds like a tiny amount with our modern diets, it is in fact not that low if you eat as nature intended. For example a standard tomato may contain 2.4 grams of carbohydrates, a carrot contains about 6 grams. As you can see, if you eat the right foods and keep them interesting, you will not go hungry.

If you have high levels of ketones in your body, it can cause ketoacidosis, but this is only usually for type I and II diabetics to worry about. Nutritional ketosis from dieting and exercise is quite safe and we have been doing it for millions of years.

So why is this relevant for your anti-aging toolkit? Carrying extra weight not only causes a myriad of health issues and speeds up aging, it increases your risk of disease and dying. So knowing how to drop those extra pounds and kilos is essential for living a longer healthier life. Ketosis is different than fasting, however you can instigate ketosis (fat burning mode) while fasting. For example, let's say you are performing the 5.2 fasting diet. You eat normally for five days but then eat hardly anything on two days (Tuesday and Thursday). You could easily put yourself into effective ketosis by eating less than 50 grams on Monday, Wednesday and Friday. This would not only give you the anti-aging effect of regeneration, lower blood pressure etc, but would almost be a slam dunk for your weight loss. If you are inexperienced at dieting, you might find it hard at first, but stick with it, try one day, then two days, you will get there. Research foods, find ones that fill you up but are low in carbs and calories (soups are a good start as they are mostly water).

Plant Based or Carnivore?

Well if you have been reading this book properly you already know the answer to this. The amount of scientific data that being a plant based eater will lengthen your health span and life span is frankly now overwhelming. There is no argument how nature intended us to eat. Dairy foods contain advanced glycation end products (AGE's). Cooked meats containing cancer causing Polycyclic Aromatic Hydrocarbons PAH's. Meat also contains methionine that goes on to make homocysteine, (note, high homocysteine levels induce heart attack, stroke, blood clots, etc). Meat also increases insulin growth factor 1 (IGF-1) to speed up cellular division. The jury has spoken and the verdict is in, plant based foods are your medicine.

Waistline - Predictor of Frailty Later in Life

Do you hold a bit or extra weight around the midriff? Well according to a recent study, if you are middle age, around 50, all that extra weight you are holding will make you enter frailty during old age, faster! So let me recap, holding extra weight around your stomach is shown to speed up aging. But how? This research was performed by the University of Oslo in Tromsø, Norway. This study tracked around 4500 people over the age of 45 for over twenty years. And what it found was quite remarkable. Those with a high or relatively high waistline perimeter were 57% more likely to be frail than their slim counterparts.

But what does the word frail actually mean? Well for the purposes of this study, it simply means things like weakened grip strength, slower walking speed, increased propensity for exhaustion, involuntary weight loss such as loss of muscle or bone (the very things you don't want to lose) and of course, poor physical activity. For me personally, all of those things are a disaster during aging, they are biomarkers for what I would personally consider increased biological aging. So regardless of what you consider biological aging to be those conditions I just outlined are things you want to avoid at all costs if you are interested in longevity or healthy aging.

So, let's look at some more data. Those that were obese at the beginning of the study, which is defined as a body mass index or BMI over 30 were 2.5 times more likely to be frail than those with a normal BMI in the average range of 18.5 to 24.9. That's quite stunning, because if you consider those conditions of frailty we mentioned above to be part and parcel of biological aging, then those with a BMI over thirty aged more than <u>two point five times **faster**</u> than their slim counterparts.

There are of course numerous reasons why obesity could increase frailty, but more confusingly, how does obesity affect muscle strength? Well obesity is implicated in things like increased inflammation in the fat, which can induce muscle fiber damage, which results in decreased muscle strength and function. So don't think of fat and muscle as two completely separate mechanisms, because excessive fat is clearly detrimental to muscle fibers, and increased muscle mass is clearly detrimental to fat build up, as muscles require vast amounts of energy, so muscles can help clean up excess glucose and reduce fat accumulation. You have to decide which mechanism is going to be the most dominant mechanism in your body.

Of course, obesity is connected to a range of other health conditions such as diabetes and even increased cancer prevalence, so changing your mindset and learning new healthy habits, and never temporary dieting, but learning a new lifestyle, is going to deliver great benefit to your overall health.

This study appears to indicate that your health at middle age can dictate your rate of biological aging later in life, or at least dictate your ability to function vibrantly, which I consider an important indicator of biological aging. I would also argue this effect is reversible with losing that extra fat and lifting weights to restore muscle strength.

This is why maintaining a clever diet, that isn't loaded with carbohydrates or a diet that makes your body hold water is a good decision. It is also why the older you get, the more you should consider some type of training.

Plant Based Foods - The Ultimate Life Extending Diet

Of course it is not just animal based products that will kill you faster, sugars disguised in the form of carbohydrates (breakfast cereals, pasta, potato, rice, bread, cakes, alcohol, etc) which turn to sugar once metabolized cause glycation, which again is the speeding up of cellular aging. Once your cells are old and dysfunctional, you become frail, slow, fragile and completely vulnerable to disease. So keep those cells young for as long as possible (have limited exposure to the foods above).

When we think of plant based diets we tend to think of raw broccoli for dinner, or cabbage leaf for breakfast, but it could not be further from the truth. The options are incredible and the range of food never gets boring. Plant based diets are one of the easiest changes you can bring about. Give yourself a month or two to adjust whilst you find the meals and foods you love.

Milks? Use rice, coconut, almond, soy, oat, walnut milk to name a few. Experiment and find the ones you love. Ween yourself off "baby bovine growth fluid" (cow's milk). This fluid is filled with hormones and lipids designed to make a baby cow grow extremely quickly and it's certainly not designed for human consumption. Cow's milk raises your IGF-1 levels which speeds up aging. Yes we have all been brainwashed to think drinking baby calf growth fluid is normal, but really it is not (think about it). In fact no species drinks milk after a few months of life, except humans that have fallen victim to the food industry's marketing.

And when it comes to the food industry brainwashing us, the same goes for many other foods such as dairy, meats and even fish, but this is an entirely different subject. Diets high in animal protein speed up aging by increasing your IGF-1 levels. Ultimately though a plant based diet is very low in sugars and animal proteins. It gets you close to calorie restriction without ever going hungry as well. It is obviously much healthier just because of the type of foods you are eating. By consuming less calories the weight will fall off you and you'll also be healthier and lighter because of it.

Get creative. I am vegetarian now, but was once like you, I could never be a vegetarian, but I cannot believe how easy it was and the rewards it has brought. For the misguided readers amongst you, there is nothing masculine about eating meat, there is only educated and uneducated, or wanting to live longer or not having the will power to make changes to live longer, or brainwashed by the food industry or not being brainwashed by the food industry. Choose the category that fits you best and decide which one you are.

Not Convinced by The Overwhelming Body of Evidence?

A quick internet search for women (mothers) that look younger than their offspring reveals some incredible evidence. There are many intriguing cases to find online about why some mothers can look younger than their offspring, the scientific clues are there for those of us that are willing to look for them.

Most of these mothers are vegetarians, and because vegetables are mostly water, this can be interpreted as a form of calorie restriction (which we know extends life). The vegetables will fill you up, and if prepared well can easily become your favourite meals.
One case of where a mum looks younger than her son is Shimmi Munshi. At 41, Shimmi still gets asked for ID when trying to buy alcohol. Shimmi does not exercise fanatically or stick to any particular training regime and when asked why she looks so young, Shimmi will tell you that she has no secret and no idea. However I know her secret and I also believe that even Shimmi does not actually realize what her secret anti-aging weapon even is.

So what is her secret (as you, the reader lick your lips in anticipation of being asked for ID at liquor shops)? How could her stem cells be so effective at repairing her appearance?
As I researched more about Shimmi's case, it became evident that Shimmi has a disease known as gastroparesis. This disease can make Shimmi vomit multiple times per day, and this is where the anti-aging science happens (for those in the know).

Gastroparesis is where the muscles in the stomach (or nerves controlling those muscles) fail to work, so the stomach becomes incompetent at grinding foods properly. There are treatments for this and recommended foods to eat (soft foods and liquids), but ultimately the science you want to know is how can you replicate this effect without initiating a gag reflex multiple times per day. Well gastroparesis is causing Shimmi's body to mimic calorie restriction by regurgitating most of what she eats. Her body's absorption of energy rich foods is low (due to her vomiting those foods out), so her body is constantly purging old cells (to save energy) and renewing her body with brand new cells keeping her appearance (and entire body) extremely youthful.

How can you replicate this effect? Easy, consume foods extremely low in sugar, low in carbohydrates and low in animal fats. Steer clear of foods such as sweets (chocolates, ice creams, etc), rice, bread, potatoes, pastas, breakfast cereals etc. Those types of foods are packed with energy, and only people performing vigorous sports or work need consume them.

Foods that are low in energy that can assist with mimicking calorie restriction are of course vegetables (they are mostly water and packed with the essentials of life). Vegetables will fill you up, keep your insulin levels low and your glucagon levels high. We are designed to eat foods low in sugar (energy) such as vegetables, and not to be brainwashed by the food industry's marketing, and Shimmi is clear proof of that.

Time To Evolve – Kill The Creature Of Habit

We were all brought up to eat foods that were bad for us. I get it, you love a steak, you love a burger, you love pasta, French fries etc. We are all set in our ways and evolving and learning new skills for many can be an effort. Who wants to learn how to cook new healthy low sugar meals when we already know how to prepare umpteen comfort meals?

Your primary evolution here is reinventing how you cook and what foods to cook. However it is really easy to prepare low sugar foods (vegetables) and set them up so that you are not only healthier and slowing down the aging process, but that you cannot wait to get home and cook them. It is quite liberating being able to make foods likes this (that you most likely do not even realize). Unfortunately, this is not a cooking book, so I urge you to research online how to prepare plant based foods that you will love. Visit some vegetarian restaurants for ideas and to relearn how to eat natural and healthy non-processed foods

Obesity Means More Cancer, More Heart Disease

Vegetables are mostly water; however, they will also fill you up preventing you from overeating. Vegetables are also low on sugars, so they will not spike your insulin level (no sugar spikes in the blood), preventing you from being hungry again (sugar spikes cause hunger).

Insulin needs to be low in order to lose weight. Insulin is the hormone that tells your body to eat more food and lock that food away as stored energy (fat). Glucagon is the hormone that tells your body the opposite, to use stored fat as energy (burning fat). Keep your insulin levels low by eating foods very low in sugar (that means no breakfast cereals, no pasta, no potato, no rice and no bread if you are serious about not being overweight and extending life), because once those foods are metabolized by the body, they become sugar. Eat them at your peril.

So, controlling and lowering your insulin levels and pushing up your glucagon level is the goal. Obviously, you are going to consume things that will spike your insulin levels and take you out of fat burning mode, however as long as these are short periods and your glucagon levels are good, your body will burn more fat and get your physique into a healthy and life extending position.

When you stop eating sugar rich foods or foods that metabolize into sugars such as –

- Pasta, bread, breakfast cereals, rice, potatoes (starch)
- Fruit juices or soft drinks
- Sweets such as cakes and chocolates

Your taste buds will become hypersensitive to sugars. Personally, I can no longer handle condiments or products such as sweet and sour sauce, rice or bread (the taste has now become sickly sweet). This is how we evolved, but your sugary diet has blinded your taste buds from being able to detect the amount of sugar you are actually consuming. By changing your taste buds to being hypersensitive to sugars, now foods you once called boring, now taste sweet and great, just as nature intended, but you'll need to get there first by changing your diet. A carrot can taste as sweet as chocolate, but your taste spuds are currently blind because of glucose saturation.

Flexitarians & Chill Vegetarians

This is not a competition about willpower or who has the most rigid diet. This section is simply showing you that a plant based diet is going to lower your blood pressure, lower your cholesterol, lower your weight and give you a ton of health benefits that will ultimately "**extend your life**".

So when avoiding dead meat, there are many ways to skin a cat (pun intended). There is a growing trend to now be a flexitarian as opposed to a militant vegan. Try and eat plant based foods most of the time, but do not feel you are competing for a gold medal. Sometimes you are going to be out and the menu will be animal products. Sometimes you will be at a family dinner and there will be animal products. Just eat plants most other times.

The point is that being a plant based eater all the time can sometimes be impossible with the food industry's marketing being rammed at the general population. Try being a 'vegetarian before six' by simply eating plant based foods before dinner time or use the eighty twenty rule; 80% of the time eat plant based foods.

It is very interesting to me that many people of today's era think that being plant based is some type of unusual diet. In fact, our ancestors for tens of millions of years ate vegan. It is our modern diet that is completely alien and poisonous to our bodies. Bacon, breakfast cereals with baby cow growth fluid (milk), sausages, hamburgers are all part of the unusual diet that humans have now normalized.

A great video easily found online called The Last Heart Attack by Dr Sanjay Gupta, a CNN medical journalist, proves (if you believe the evidence) that a plant based diet will make you "heart attack proof". Yes, heart attacks are 100% preventable with a plant based diet. Imagine that, we could see the last heart attack in the human race this year if everyone went plant based. The evidence is pretty incredible. The only questions are, how long do you want to live? How important is your health? Is taste more important than being disease free?

Let's just pause for a second and absorb what was said in the last paragraph. That is, that heart disease is 100% caused by diet. Yes, that's right, we could live in a world where there were no more heart attacks. Every heart attack (if you believe the evidence) is caused by the food choice's we make. Again, this is incredible science and knowledge. You now hold in your hand absolute power in guaranteeing that you never die of heart disease or a heart attack. Just think about that for a second, before you picked up this book, heart disease or heart attack was still one of the primary ways you were probably going to die (and prematurely). Now with a vegetarian diet combined with some very basic self-control, you can totally disregard that risk of disease and death.

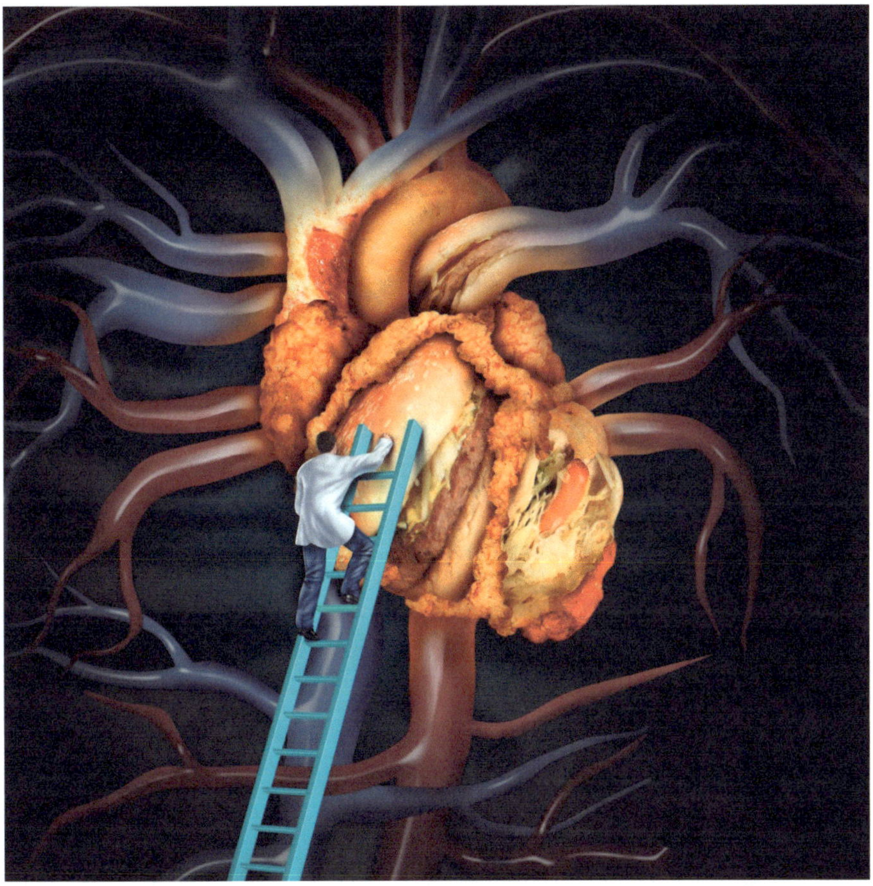

Make sure you spread that knowledge to your friends and loved ones. Watch the video by Dr Sanjay Gupta online sometime as well. However Dr. Caldwell Esselstyn is featured in that video and is the scientific guru behind the research and story.

One of the fastest ways to age and develop disease is to carry excessive weight. I want you lean and mean, and this physical state is very easy to achieve by simply changing the majority of your dietary intake into vegetarianism.

Late Night Eating Habits

Everyone knows that eating late at night is a recipe for weight gain and a slower metabolism. but recent data shows even more reason why that late night packet of potato chips and 11pm frozen pizza is going to kill you faster.

A study by researchers at Brigham and Women's Hospital analyzed why eating late at night as opposed to regular eating habits had such profound health differences. The research showed that eating late at night increased risk for obesity, increased formation of fat tissue and decreased ability for weight loss.

For the study, overweight / obese people were selected for two in-laboratory eating schedules. this was a crossover study, which simply means the group of 16 people that were selected all receive the same treatment schedule, but the order in which they receive them is different and assigned randomly. Facets such as sleep, light exposure, physical activity and even posture were all taken into account.

Both eating schedules delivered exactly the same foods, but one was given to the subjects approximately four hours later in the day, pushing the calorie intake into the evening.

Regular blood tests were taken during this trial, including things like their hunger, body temperature, energy expenditure were all observed. Biopsies were also taken of fat tissue in order to monitor various biochemical behavior in relation to adipose tissue formation or basically fat accumulation. Needless to say, the results between the two different timed diets were quite fascinating.

Consuming foods later transformed or changed the behaviour of the appetite regulating hormone ghrelin which resulted in lower levels of the hormone leptin. Leptin basically instructs your mind to tell you that you are full and to stop eating. So, what do you think happens when these levels are lower? When people ate later, not only did their appetite hormone ghrelin kick in more making them hungry, but the leptin levels were reduced, instructing them to keep eating. Combine this with the fact that eating later slowed down calorie burning, and we have a recipe, pun intended, for weighing scale disaster. But there is more bad news, gene expression in those fat tissues had changed, causing their tissues to prefer the storing of fats over burning their existing fat supplies in those tissues. So eating late at night is a bad idea for anyone trying to lose weight or stay healthy.

I personally don't eat till lunchtime, and then I stop eating at 6pm. That is an 18-hour window each day where I do not eat, or a six hour window from lunchtime to six pm, where I do eat. But everybody is different, what works for me, may not work for you. Your age, sex, medical conditions, medications, weight, even ethnicity can change various facets of your biology.

Many people struggle with weight, and there are many tactics you can use with the right information and the right mentoring. I am yet to find someone that couldn't be transformed.

Don't fall into the trap of using your sense of taste to dictate what foods you eat, that is a pathway to having poor health. Learn to eat foods that may not taste nice but are full of nutrients. Your mind should be making food decisions as opposed to your sense of taste. And I'll say it again, your taste buds do change over time, so foods you may not like, become quite easy to eat after a short time.

Semaglutide / Liraglutide / Tirzepatide- Weight Loss and Control

There is a weight loss section at the back of this book. However, I think it is worth including three pharmaceutical weight loss drugs here called semaglutide, liraglutide and tirzepatide.

Semaglutide works by mirroring the action of Glucagon-like Peptide-1 (GLP-1), which releases more insulin into the blood, which lowers levels of blood sugar. Liraglutide helps release insulin after eating. Both semaglutide and liraglutide work effectively in weight management and weight loss. Semaglutide seems to be a little more potent for weight loss, however both are very effective and dosing size can account for the differences. The difference during a clinical trial showed that semaglutide caused weight loss of around 15% compared with liraglutide at around 5-6%, and again, you could make up for that difference by changing the dosing, but this is something you would need to chat with your doctor about. Tirzepatide beat Semaglutide in weight loss with a trial showing 17.8% to 12.4%. All three work and slightly overweight people do not need the most powerful.

Note you may think that 15% or 5-6% is quite a small amount of weight loss, but these are morbidly obese people, so the weight loss was quite vast. Intended for diabetics and the morbidly obese, these drugs have now hit the mainstream market, where people with relatively good physiques are now using them for that perfect social media physique. Even though they are not performance enhancing drugs, they are certainly carving out their own market. In Australia, these drugs are very difficult to get, as the manufacturers cannot keep up with demand.

But there is some caution to take heed of. You may not feel hungry on these drugs, in fact you may feel full for most of the day, so you won't really need will power to stop eating, as the thought of eating may give you nausea (which is one of the side effects). You will certainly lose weight when you stop eating, but in some studies the weight loss was mostly muscle and other tissues instead of fat. This should tell you that you need to lift weights, eat high quality amino acids or proteins during this time and only this way will the fat disappear whilst the healthy tissues remain. As a scientist that studies aging, I cannot stress enough, you do not want to lose muscle mass or other tissues (this is a fast track to frailty). The fact you are reading my book is great, so lift weights and keep that muscle on, especially if you are using GLP-1 agonists like the above.

These are all pharmaceutical drugs and they do come with other side effects not mentioned here, so you will need to chat with your doctor regarding which one is best for you and the relevant dose.

All of these GLP-1 receptor agonists are found under names such as Wegovy, Ozempic (semaglutide), and Saxenda (liraglutide) and Mounjaro (Tirzepatide).

If weight loss interests you, then make sure you also read the weight loss section at the back of this book, because you will gain an understanding of the basic biochemistry at play when it comes to storing and losing fat. It is quite simple once you understand a few rules.

Ref: Effect of semaglutide and liraglutide in individuals with obesity or overweight without diabetes: a systematic review. Ther Adv Chronic Dis. 2022 Jul 4;13
Ref: Efficacy and Safety of Liraglutide and Semaglutide on Weight Loss in People with Obesity or Overweight: A Systematic Review. Clin Epidemiol. 2022 Dec 6;14
Ref: Effect of Weekly Subcutaneous Semaglutide vs Daily Liraglutide on Body Weight in Adults With Overweight or Obesity Without Diabetes: The STEP 8 Randomized Clinical Trial. JAMA. 2022 Jan 11;327(2)

Peptides

Peptides are short chain sequences of amino acids (usually under 100 amino acids). The sequence of amino acids can code for small proteins or emulate a key going into a lock to induce the release of hormones or some other type of biological function. Peptides occur naturally in our bodies, and are also synthetically made, and can be used for a range of purposes. The three weight loss drugs mentioned on the previous page are all peptides. I personally do not use peptides, but they are growing in popularity and will be important for future health and longevity as science becomes more sophisticated, so peptides have made it into my book. So here is a list of peptides worth a mention –

CJC-1295 / Ipamorelin is a growth hormone secretagogue, which is a amalgamation of two peptides often used together to boost production of growth hormone. CJC-1295 is a growth hormone releasing hormone (GHRH) equivalent which acts on the pituitary gland to increase the discharge of growth hormone. Whereas ipamorelin is a growth hormone releasing peptide that activates the ghrelin receptor at the pituitary gland to block the inhibitory effect of somatostatin (a growth hormone inhibitor) and further increase growth hormone secretion.

Amlexanox inhibits histamine and leukotrienes and can improve glucose management leading to decreased weight, insulin resistance, fatty liver and inflammation. Amlexanox has antioxidant, anti-inflammatory and immunity-improving assets, but interestingly can trigger AMP-kinase (AMPK), which enhances metabolism and longevity. It can also encourage cells to burn fat as a fuel source, controls appetite and promotes weight loss, improves Hemoglobin A1c, insulin sensitivity and fatty liver.

MOTS-C is a mitochondrial peptide that can foster mitochondrial biogenesis, enhance the ability of mitochondria to burn fat, increases healthy brown fat and activates AMPK which enhances glucose metabolism. MOTS-C also improves insulin sensitivity and glucose metabolism through encouraging glucose uptake into muscle cells. You could literally regard MOTS-C as an exercise-mimetic, meaning it emulates exercise on the body, advances weight loss, heart health and bone formation.

5-AMINO-1MQ can improve cellular metabolism (converting fat into energy) and enhance mitochondrial function, whilst at the same time appears to increase NAD+ levels, enhances weight loss and can increase muscle growth. It may also have benefits in reversing type II diabetes and diet-induced obesity.

SS-31 is regarded as a cellular anti-aging peptide, which protects mitochondria against free radicals and boosts mitochondrial function and ATP production, whilst also targeting senescent cells (which you will read about later).

Epitalon could be the ultimate longevity peptide for your anti-aging toolkit. Demonstrated to increase telomere length via amplified telomerase activity and reduce mortality in animal models, epitalon can also induce serotonin and melatonin and has also shown to slow down aging of the reproductive system. Epitalon increases conversion of T4 to T3 thyroid hormone (normalizes deiodinase function) and improves glucose tolerance and decreases insulin and triglyceride levels. Studies have shown reliable increases in longevity and a reduction in mortality and cancer. We also see improved hypothalamic-pituitary dysfunction seen across some illnesses and aging. Epitalon can deliver immune support by improving natural killer cell and T-helper cell function.

- Be super careful where you purchase from, and always chat with your doctor first.

Neurological Health or Decline

Short bursts of ketosis are beneficial for the brain and so is fasting. Have your last meal at seven o'clock at night, then your next meal at least 15 hours later, lunchtime would be even better. Maintain low sugar diets to ward off neurological disease as well. And remember, sugar is bread, pasta, potato, rice, pastry and any other heavy carbohydrate foods.

Of all the things humans can do to ensure that their neurological health remains optimal, exercise is most powerful thing of all to ward off diseases such as Alzheimer's. Exercise assists with clearing out beta amyloid plaques and maintains healthy blood flow throughout the brain. As already discussed earlier in this book, combining different types of exercise, such as aerobic exercise, strength training exercise and flexibility exercise will get the best results. Try and exercise a minimum of 4 days per week and aim to get your resting heart rate below sixty.

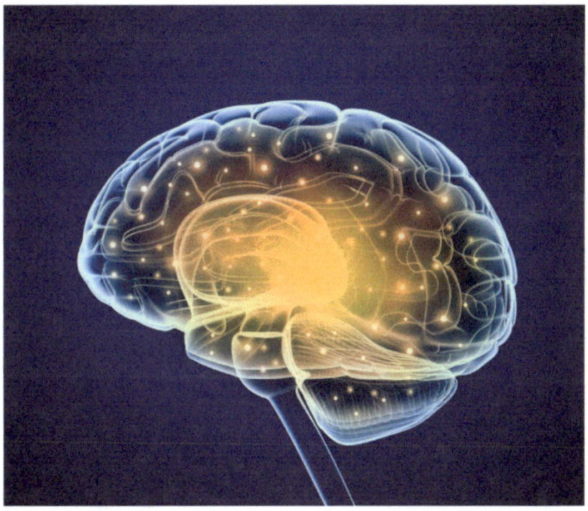

A 2017 study published in the journal Frontiers in Pharmacology showed that cannabinoids appear to assist with long term memory. And when mixed with THC (the active ingredient inside cannabis), the effect on memory was even greater. This is in stark contrast to what is commonly believed about those that inhale this green whacky weed. In fact cannabinoids remain one of the best treatments to ward off neurological diseases "as well as Parkinson's and epilepsy". I certainly do not use any type of cannabinoid, but the fact this powerful agent has been around for millennia and is only just being accepted shows just how far behind our mainstream medical thinking is (so remember that fact if your doctor thinks your quest for longevity science is nonsense). There are many ways to ingest cannabinoids these days without smoking or getting high.

Some oils also appear to show a therapeutic effect on neuronal health as well and namely coconut oil ranks high in this debate. Of course if used in excess all oils may cause cardiovascular problems, so each person needs to weigh up their own dosage and requirement (if any) under their doctor's supervision.

And remember fasting is also proven to induce neurogenesis (the creation of new brain cells), whereas binge eating has the opposite effect.

Metformin is also a great neuro-protector of your grey matter.

Alcohol – Formally Ethanol

Yes, this topic has to be covered. Most people we know drink this stuff. Alcohol is a Class 1 carcinogenic (according to the International Agency for Research on Cancer / IARC). In fact the IARC considers at least 3.6% of cancer cases worldwide are attributable to alcohol.

How can you slow down the damage it causes without abstaining?

Firstly, if you know you are going to drink, eat food. The more you plan to drink, the more you should eat. The only time I usually eat pastas or breads is when I know I will be at a wedding or event where alcohol could be served for an extended period. I eat these foods because they prevent the alcohol moving directly to the liver and assist with the alcohol being broken up in the stomach (which reduces stress on the liver).

- So to you young men out there, have the kebab at 7pm, not 3am.

Drink low carb beers and go easy on the spirits. Of course wines are a popular choice as well.

When it comes to reds, I drink Pinot Noir's because they are lighter and apparently contain more resveratrol than other reds (but the resveratrol content is irrelevant). When it comes to whites, I have not found any science to justify any type of white wine over another (except sweeter whites contain more sugar).

As you would have read earlier in this book, NAD is a co-enzyme that is required to maintain another enzyme called PARP1. PARP1 is a repair mechanism inside the nucleus that repairs DNA. So keep your NAD levels high when drinking alcohol.

Also use curcumin when drinking alcohol because it is a great little epigenetic regulator, and can push back against the harmful epigenetic changes that alcohol wants to cause. Same with green tea and folate, both great epigenetic regulators.

Try not to drink every day. I personally abstain from Monday to Friday when possible.

If I am going to be in an environment where there will be extended consumption of alcohol, I always have electrolytes with me (sugar free of course).

The hangover is caused by two things; Dehydration and lack of electrolytes.

- Contrary to what you may think, you cannot hydrate with just water. You need electrolytes for the water to be absorbed.

Usually these electrolytes come from the food you eat, however the process could be slow depending on your body. Drinking electrolytes with water speeds up the recovery and repair process.

Cold water cannot be absorbed by the body either, so if you are feeling hungover (dehydrated) from the night before, then drink warm water with electrolytes and your hangover will disappear much faster.

And yes, I am that much of a strategist, I will take electrolytes before, during and after any extended alcohol consumption. And it is okay to keep eating throughout that experience as well, as that will keep the pressure off your liver. I rarely travel or leave home without my electrolyte stash.

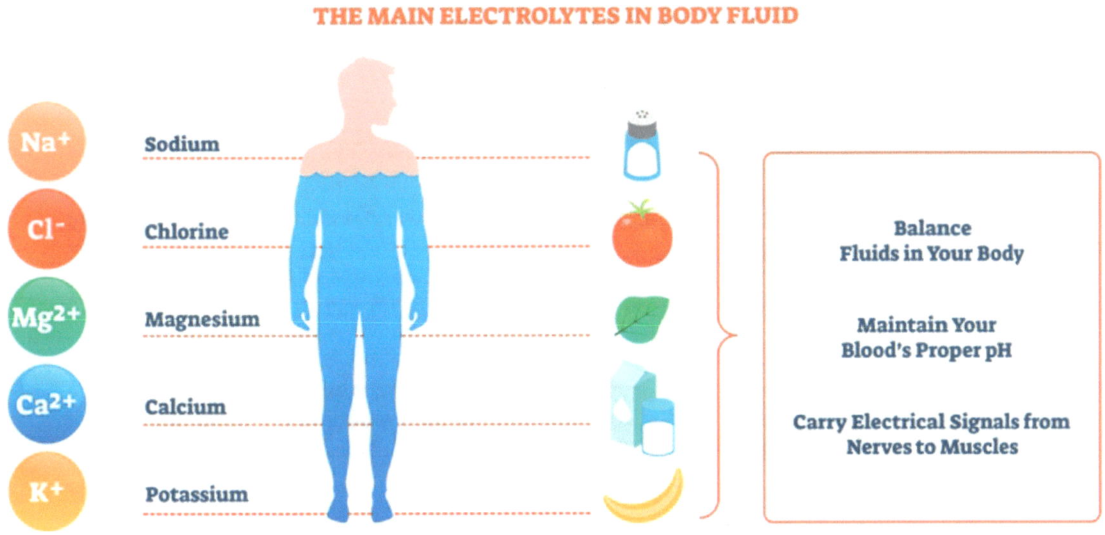

Note I always have zero alcohol beers in my fridge. These are relatively new on the market, a healthier alternative and great to have in-between other drinks to slow down the effects of alcohol on the body. Of course a glass of water doesn't hurt during the soiree. Low carb or zero carb beers are also highly recommended.

Metformin also slows down the release of sugars from the foods you eat; and alcohol is full of sugar. So metformin is also another good defense to alcohol consumption.

A study from Boulder University, Colorado found that exercise could also reverse damage to the brain that was caused by alcohol. Just light exercise such as riding a bike showed improvements to reverse the damage.

Hydration and Lifespan

A recent study published in the journal eBioMedicine infers that people who don't drink enough fluids may be at a higher risk of chronic disease and are shown to die at a younger age. How much younger you ask? Well a 2019 study that deprived mice of staying properly hydrated, found that the mice died up to six months earlier, which is about fifteen years earlier for humans. So in this latest study, done by the same team that performed the 2019 study, the researchers looked at an ongoing heart health study that started back in the 1980's and encompasses over 15,000 subjects.

So what did the researchers look at? Well they looked at sodium levels in blood samples because this is a good indicator of hydration. If you are healthy, your sodium levels are usually between 135 millimoles and 146 millimoles per liter of blood. So this sodium in the blood, which I will refer to as serum sodium was measured multiple times over 25 years in these subjects, in parallel with other health and aging bio-markers such as blood sugar levels, blood pressure, immune molecules, etc.

Before we continue, let's take a quick look at what sodium is. So for those of you that don't know, table salt is sodium chloride, with a 1 to 1 ratio of sodium and chloride. And sodium alone is an akalai metal and is the sixth most abundant element in Earth's crust.

So what does this mean for you? Well in biochemistry, higher levels of something doesn't necessarily mean healthier, and lower levels certainly doesn't mean healthier either. It appears with most vitamins, minerals and other compounds essential for life, there is a range or sweet spot that you need to aim for, and these ranges or sweet spots can change with age.

So with this study, the scientists found a connection with serum sodium levels that were higher than 142 millimoles per liter and faster biological aging. Wow, but by how much you ask? Subjects in this study with sodium levels higher than 142 millimoles per liter, were associated with a 39% increase risk to develop chronic disease, and this ratio spiked to 50% in those with serum sodium levels above 144 millimoles per liter. Holy moly!

So when looking in parallel at their disease risk and higher sodium serum levels what did they find? I'll quote directly here "The analysis demonstrated that middle-age serum sodium in the upper part of normal range (135–146 millimoles per liter is able to predict a faster rate of biological aging" end quote, Wow!

Its worth noting, that poor diet or exercise can also have similar consequences, so hydration is only one risk factor, but this clearly adds to your rate of aging and disease potential. There are three things you won't find in my home, and that's salt, sugar or ashtrays. Those things are from a bygone era, we know much better now that these things will kill you faster, and those of you that are serious about longevity should rethink having those in your homes. So aim to change your behavior on emerging data, it is a growth and an empowering mindset.

Some of you may think you need to add salt to your foods for good health, but there was a reason I deviated into geology earlier and mentioned that sodium is the sixth most abundant element in the Earth's crust, and that's because sodium is found everywhere, and in pretty much every food throughout the human diet, so you are already getting sodium whether you want to or not.

Salt is essential for life, I am not saying it will kill you, I am simply showing you data, that elevated levels increase your rate of aging (according to the study) and also increase your propensity for disease.

So what are the best ways to lower sodium levels? For a start, stop eating processed foods, they are often packed with sodium, even when they say things like "low salt" or 50% less sodium, there is actually no metric they are being held to, so when you see wording on packaging like "50% less sodium", well 50% less sodium than what? These are just advertising gimmicks to make you think you are eating healthy foods. So learn to read the nutritional facts on the back of packaged foods, it won't take you long to learn which loaf of bread or other packaged food to put back on the shelf.

Those of you that eat in restaurants should also ask for no salt to be added to your meals. Chef's often have this request and are more than happy to offer you this healthier option. And there are many other spices you can use to spice up your foods as opposed to salt.

Foods high in potassium such as avocados or bananas will lower your serum sodium levels as well.

And as we just learnt, alcohol also dehydrates you as well, so make sure to watch your consumption and consider taking electrolytes if you are drinking for a prolonged period, as the potassium in those electrolytes will bring down your serum sodium levels and rehydrate you faster.

And keep this potassium sodium feedback loop in mind, for when you eat fast food such as a ton of salty chips, because the potassium in other foods lowers serum sodium levels. So if you eat salty junk food, make sure your next meal is high in potassium, such as an avocado, banana, spinach, broccoli, etc, because the potassium will purge much of that sodium out through the urine. This is biohacking 101.

We are all going to consume salt at some stage. And yes, I'm not saint, I'll be in a restaurant where I am consuming bread with balsamic vinegar and salt and I will thoroughly enjoy it, but it's not every day, and I'll purge that sodium with a high potassium meal later in the day.

And your other option is exercise to purge sodium. Exercise causes you to lose water and salt, which means when you rehydrate with just the water, your sodium levels are back to baseline where they should be.

You wouldn't think a simple condiment sitting innocently on your dining table was a silent killer or a protagonist of accelerated aging, but the data appears to show that it is. So, grab your saltshaker and stick it in the bin, you'll make yourself more disease resistant and be doing your loved ones a favor (much to their horror). When the kids or your partner go into full meltdown, just blame it on me if you need to, I'm happy to help. □

Acid and Alkaline - The Body's PH Levels

Acetic acid can lower the output of insulin. Acetic acid also improves insulin resistance. Insulin is the hormone that tells your body to store food you have just eaten as fat. Insulin is also the hormone that tells your body not to use stored fat reserves for energy. Ultimately insulin causes you to gain weight and is the hormone most people want to control for a healthy weight.

Apple cider vinegar is basically acetic acid and can be consumed as a liquid or even a pill. The best time to take apple cider vinegar is in the morning and again in the evening, possibly with a meal.

Even though apple cider vinegar does not have any direct life extending properties, it still gets a mention in this book because it assists with balancing your blood PH levels and lowers insulin. Let's look at some of the common age related conditions that can occur in people over 50.

The PH of acetic acid is around 2.5 (this is close to the stomach acid PH levels). You can take acetic acid for heartburn / acid reflux because as you age your stomach becomes more alkaline. To balance the alkaline, you need more acid after 50 to be able to absorb minerals such as calcium and even digest protein. When your PH level is too alkaline, you can no longer absorb calcium or minerals as effectively as you once did.

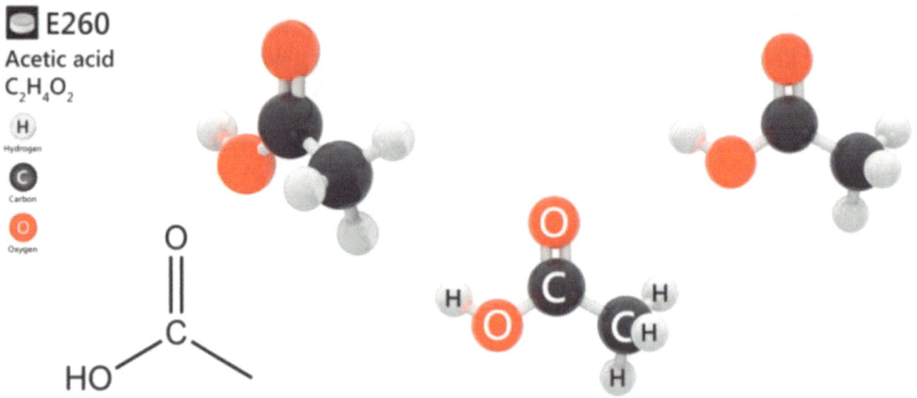

Older people develop more kidney stones, soft tissue calcification, cataracts, more arthritis, gall stones, bursitis, tartare on the teeth, body stiffening, their skin is thickening; this is calcification and may be (and usually is) the result of your body's alkaline level's being too high (alkalosis). These are age related conditions and the magic pill (if it's alkalosis causing those conditions) is maintaining a healthy PH state with apple cider vinegar (acetic acid).

Guanidine is a byproduct from eating too much protein. Guanidine is very alkaline and can cause nerve pain, arthritis, eye twitching, etc, so acetic acid / apple cider vinegar is a great maintenance tool for keeping your body younger.

Because acetic acid can assist with lowering your insulin output, taking it means you are less likely to store food as fat (put weight on) and you are more likely to use fat reserves as energy (burn fat). Apple cider vinegar (as a pill or liquid) should become part of your anti-aging toolkit.

We briefly touched on alkalosis on the previous page and what conditions may come from that -

- kidney stones
- soft tissue calcification
- cataracts
- arthritis
- gall stones
- bursitis
- tartare on the teeth
- stiff body
- skin thickening

We've also talked about the stress hormone cortisol. Guess what hormone promotes alkalosis? Cortisol does. So unless those age related conditions above appeal to you, this is all the more reason to ensure stressful people and environments are no longer part of your life. Alkalosis and cortisol speed up aging. Being stress free and less alkaline are two powerful ways to fend of aging.

This book is intended to focus on hardcore science. However because what you put in your mouth influences the science of your body so much, we cannot teach or show you how to stay young without very briefly covering the topics of weight management and diet.

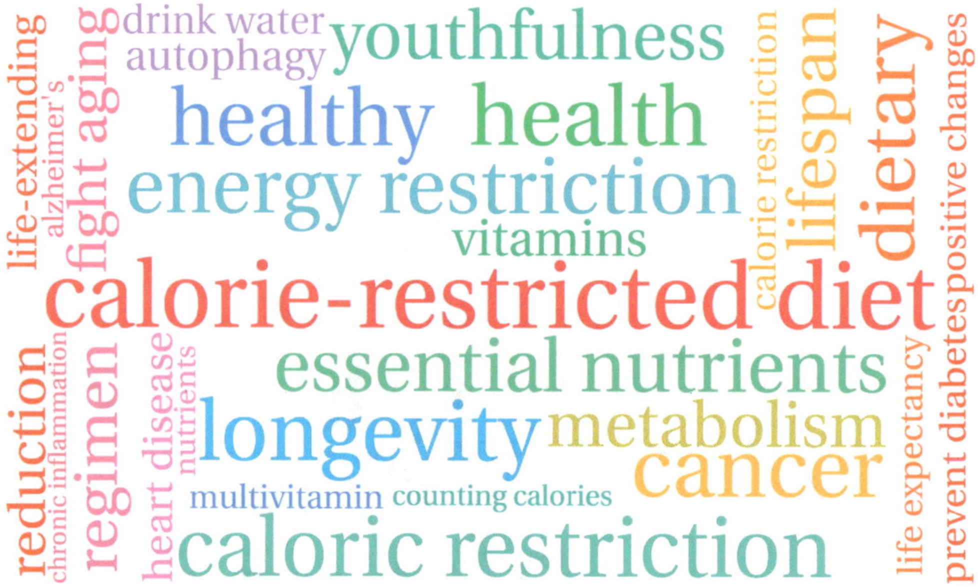

- Weight loss is 95% done in the kitchen by simply selecting the right foods and portions to eat.
- And extreme longevity is achieved by shutting down pathways to aging and disease by combining the products and knowledge shown throughout this book.

We have entered an era in human evolution where we can now control the incredibly complicated science that causes us to age. So why aren't pharmaceutical companies and governments teaching us this knowledge at school or on television then?

- Pharmaceutical companies do not make money by keeping you young and healthy.
- Many (not all) people in positions of power have very little knowledge about longevity science. New concepts that do not bring votes are rarely political or election issues, which is why progress can be so slow in governments. We need to constantly self-educate, period.

Amino Acids Worth a Mention

Acetyl L Carnitine

Anti-Aging Score 3/10
Acetyl L Carnitine is an amino acid that should reside in your anti-aging toolkit. This interesting amino acid (like many of the products we focus on here) assists with energy production in the mitochondria, the cell's battery. Note that Acetyl L Carnitine has a synergistic effect when taken with CoQ10. The other primary reason for taking Acetyl L Carnitine is because there is evidence that it can slow or stop the formation of advanced glycation end products (AGEs).

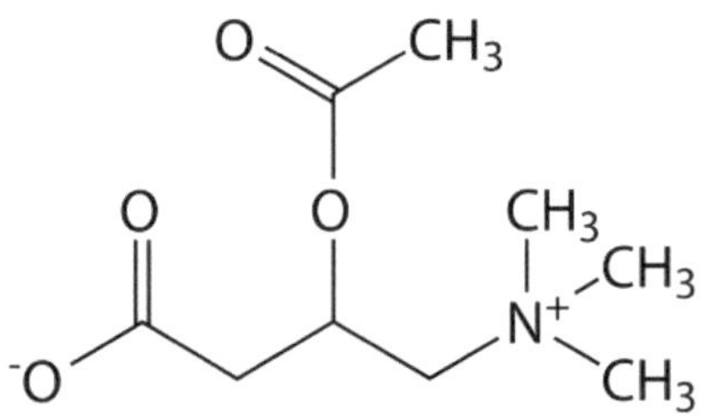

L Arginine (nitric oxide) Arterial Dilation / Heart Disease

Anti-Aging Score 3/10
L Arginine's most important task is to increase nitric oxide inside the human body. Nitric Oxide is important for arterial dilation (better blood flow).

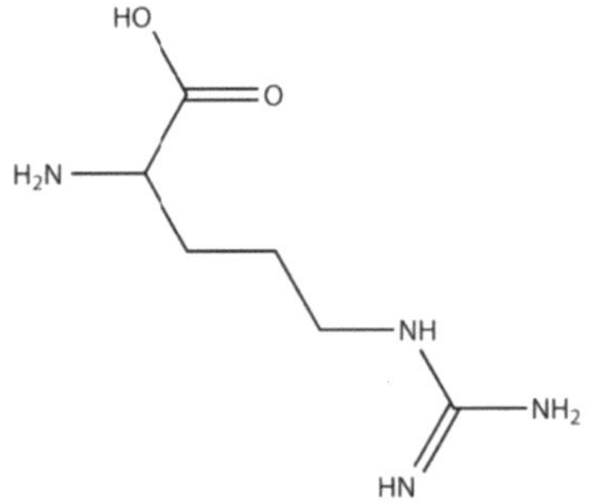

Nitric oxide is important for opening up the arteries. This can reduce blood pressure, chances of heart attack and even improve sexual health in men (better blood flow). There are many other uses for this amino acid but it has landed in your anti-aging toolkit for one reason alone and that is its ability to relax arteries and ensure better blood flow. Stiffening of the arteries is a major cause of death and disease, so arterial health is paramount if you are serious about longevity. I would advise that you consider this supplement if you eat meat or are sedentary.

Appl Physiol Nutr Metab. 2012 Feb;37(1):115-26.
doi: 10.1139/h11-144. Epub 2012 Jan 17.

Glutathione (do not use directly)

Anti-Aging Score 8.5/10

Glutathione is arguably the most potent anti-oxidant in your body, however as you age your glutathione levels drop, and as a result your exposure to oxidative stress and the signs of aging increase. This is our primary reason for placing glutathione into your anti-aging toolkit.

Glutathione can be depleted from lack of sleep, stress, certain medications, or poor diet. Glutathione is the most important antioxidant that the body can make, so ensuring optimum levels is paramount. Glutathione is made from three amino acids (glutamate cysteine and glycine) which makes it a peptide.

Traditionally glutathione can be difficult to absorb through the digestive system, as it can break down very quickly (even whilst it is sitting on the shelf as the molecule breaks down when exposed to normal atmospheric conditions and room temperature). So instead of supplementing with glutathione directly, some research shows that it may be better to supplement with the precursors of glutathione.

There are online sellers promoting specialized absorption technology called liposomes which we will look at shortly to ensure your glutathione levels are well maintained, but because these liposomes can slide past the body's biochemical checkpoints, there is a small chance that you could end up with dangerous levels of glutathione. With that said, I am yet to read about any actual incident of this happening, but its noted.

Glutathione is found in all cells throughout your body to remove the free radicals that could be speeding up age related decline. As cells process energy, they release what are known as "reactive oxygen species or ROS". This is just a fancy way of saying "free radicals" (among others), and glutathione is extremely effective at removing these products that may cause oxidative stress, damage, disease, or inflammation.

When taking the building blocks for glutathione, and in this case Glycine and Cystiene or NAC (N-acetylcysteine), glutathione levels were restored back to youthful levels which improved levels of oxidative stress, mitochondrial dysfunction, inflammation, insulin resistance, endothelial dysfunction, genotoxicity, muscle strength and cognition. So, glutathione ranks high on my anti-aging scale, but I would urge you to use Glycine and N-acetylcysteine as opposed to glutathione directly.

Dosing was glycine (1.33 mmol/kg/day), N-acetylcysteine (0.81 mmol/kg/day). Just times your weight in kilos by the (mmol) millimole amount. Then use an online millimole converter to work out how many milligrams you need.

You can also switch out glycine for serine according to another paper.

Ref : Results of a pilot clinical trial. Clin Transl Med. 2021 Mar;11(3):e372. doi: 10.1002/ctm2.372.
Ref: Combined Metabolic Activators Accelerates Recovery in Mild-to-Moderate COVID-19. Adv Sci (Weinh). 2021 Sep;8(17):e2101222
Ref: A comparative crossover study. Redox Biol. 2015 Dec;6:198-205. doi: 10.1016/j.redox.2015.07.012.

N-Acetyl Cysteine aka NAC

Anti-Aging Score 8.5/10

Let's talk more about NAC. Cysteine is a non-essential amino acid, and N-Acetyl Cysteine is simply a stable form of Cysteine. Because it raises glutathione, it can assit with things like alcohol toxicity, the side effects of some medications, reduce scarring from acne and reduce lung issues such as pulmonary disease and mucous. NAC also reduces inflammation and has been shown to also limit liver damage. This product is important for those of you that may drink excessively (also consider an NAD precursor, curcumin and apple cider vinegar if you drink excessively).

Take it twice a day, one time in the morning and then again in the evening, as serum levels only last for around eight hours, so keeping your NAC levels higher for longer will see you remain more resistant to various forms of tissue dysfunction. Remember, aging is a slow creeping dance of decay, and molecules like NAC will prevent much of that dysfunction.

Hyaluronic Acid

Anti-Aging Score 7.5/10

Hyaluronic acid acts as a substrate for our connective tissue and also as a lubricating molecule which is a major component of synovial fluid (the lubricating fluid found in joints). Hyaluronic acid is a primary component of skin as well, and the body uses it for repairing damaged tissue. This equates to increased wound repair and tissue regeneration, response to inflammation and angiogenesis (the formation of new blood vessels).

The free radical scavenging ability of hyaluronic acid also assists in protection against excessive sunlight exposure. Hyaluronic acid also acts as an extracellular matrix and assists in tissue (skin) hydration. Cells need to move around to repair wounds or other types of damage and hyaluronic acid allows the medium for the cells to travel through. So, maintaining hyaluronic acid levels through age is extremely important and is why it made my list. Anything from sore joints to dry eyes can be caused by low levels of hyaluronic acid.

So, can this product be taken orally? And does it actually work? Yes, and yes. We have good human data that taking hyaluronic acid orally does make it through the gut into the bloodstream and does improve wrinkles. So, this little molecule is quite underrated in aging circles, because there are two schools of thought when it comes to anti-aging technologies, and that is the prevention of biological decay or beauty. I'm concerned about the former, but it's nice to know the latter (skin health) is available too. I have seen evidence that hyaluronic acid does NOT work in skin creams, and for me, that is obvious as the molecule (weight and shape) is just not going to pass into the skin effectively. Maybe I have missed some other skin cream or brand, but the science (and some studies) tells me that hyaluronic acid is not effective dermally.

There are different versions of hyaluronic acid, and I would lean toward the high molecular weight versions that are available.

Health Supplements Versus Longevity Supplements

Note, that unless you are deficient in vitamins or minerals you are wasting your hard-earned money when purchasing vitamins and minerals. A balanced diet is all you need, chat with your doctor for a blood test.

Vitamins and minerals can be great, but only if you are deficient in them. Vitamins, whey protein, nutritional supplements hold merit, but if you are taking them, and them alone, then you are still cruising to old age and a higher risk of disease at one second per second. The point of this book is to slow down the onset of age-related conditions, slow down the rate at which we age itself, and with some of the other technologies that I will introduce you to later in this book, rejuvenate those tissues to a youthful state.

Vitamins and minerals remain important for your health and well-being (in fact you'll die quite quickly without them), but those things alone will do little to extend your life, slow down aging or keep you young and vibrant. You need longevity technology and serious science to fight the process of aging and lucky for us all the technology for extreme longevity is beginning to arrive. The point of this paragraph is to reaffirm that those who argue that a balanced diet and healthy living is the key to a long life; well they may be right in one sense, however the key to a longer life and youthful living is the longevity science shown throughout this book.

When your friends and family think longevity science is nothing more than healthy living, recognize their lack of scientific knowledge and be glad you have evolved past that way of thinking. Why would anyone try and live a healthy lifestyle and learn about vitamins and minerals (which is the basic form of anti-aging and longevity science) but then fail to become literate in longevity science talked about in this book? That's like a mechanic learning how to service a car engine without knowing how to service the brakes, sure the car is going to run okay for a while longer, but ultimately it is going to end up on the scrap heap much earlier than it should.

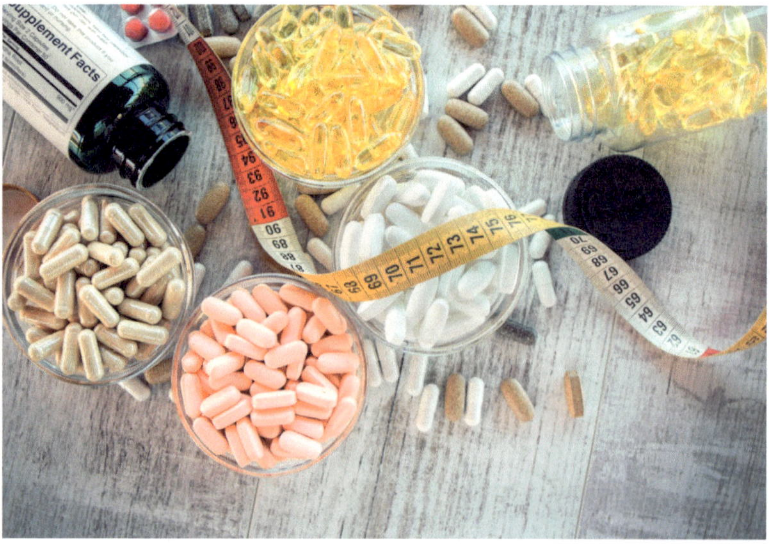

Vitamins and Minerals

Many of you reading this book may not have seen the surgical difference yet. Vitamins and minerals may give you stronger fingernails or better hair (but only if your diet is deficient). However, they cannot target longevity pathways to slow down, stop or reverse aging. Vitamins and minerals are a completely different class of supplement than what you are learning about in this book.

Vitamins & Minerals	Longevity Supplements
Fingernails	Target Aging Pathways
Healthy Aging (maybe)	Slow Down Cellular Division
Hair	Extend Cellular Limits
Teeth	Moderate Metabolic Functions
Bone Health	Clear Up DNA Mutations (anti-cancer)
Wellbeing	Targeting Cellular Machinery
Effective Only If Diet Is Deficient	Targeting Epigenetic Drift
Minimal Overlap into Longevity (maybe)	Maintain Cellular Machinery (ATP, APMK)
Etcetera	Etcetera

A study from John Hopkins Medicine (an institution that uses physicians and scientists from John Hopkins University) performed **a mass analysis of 277 published studies** and found that almost all vitamins and minerals could not support any discernable difference in lifespan (or heart disease).

If your diet is sound, then chances are you do not need vitamins and minerals sitting on your kitchen bench, so be critical of what you spend your money on. I take around 13 longevity pills per day, and you can learn about my regime later in this book.

The study from John Hopkins Medicine was called –

- **Vast majority of dietary supplements don't improve heart health or put off death, study finds**

A multi-billion-dollar industry built on the belief that vitamins and minerals will hopefully ward off disease and death. Ouch!

Bioavailability and Absorption

Ultimately you want the products I talk about in this book moving around your bloodstream as much as possible. Many health supplements that can be purchased may not even make it past your digestive wall and into your bloodstream, however with some careful science, all products can get to their intended target.

There are other ingredients that are often found mixed with the active ingredient inside your health supplements to maximize absorption. One of the most popular products used for this is piperine (found in black pepper and long pepper) which gives pepper its distinct smell. Piperine may have mild anti-inflammatory effects but alternatively may also cause adverse interactions with medications. Piperine is an alkaloid which is only really used when taking oral supplements. Putting black pepper on your healthier foods such as salads or soups can assist with taking up more of the nutrients that the food has on offer.

However there is more than one way to get your enzymes and anti-aging products into your blood stream. IV drips are popular and clinics have popped up in numerous locations across the globe to supplement you with enzymes such as NAD or resveratrol. However IV drips are expensive, time consuming (3 to 4 hours for a short treatment) and require a professional medical practitioner such as a nurse to implement them. You cannot short cut an IV drip with someone that is not trained otherwise you risk death if anaphylactic shock takes place. This means that IV drips remain expensive for most people.

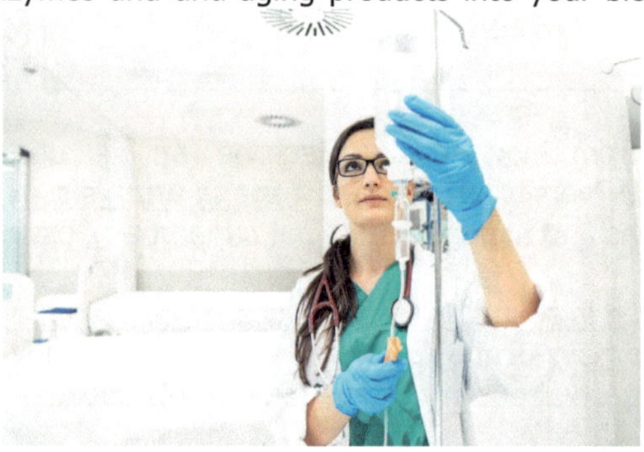

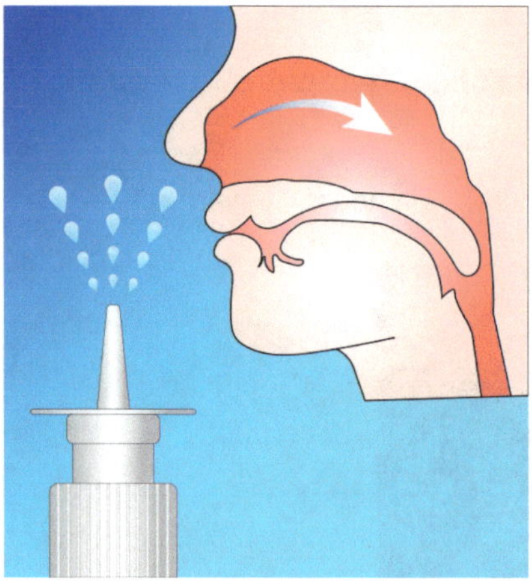

However another pathway of getting your enzymes directly into the blood stream exists, and with myself being an engineer of biotech equipment, we have the perfect device (which is close to commercialization) for you. Nasal sprays have been around for decades and the intranasal pathway allows good levels of whatever enzymes you may be spraying to enter your blood stream. Interestingly, I your trusted author have a worldwide patent on a dual chamber nasal spray which mixes a liquid with a dry molecule once the nasal spray is first pressed. Many molecules such as NAD have a relatively short shelf life once mixed into a liquid solution. Keeping the active molecule separate from the liquid extends the shelf life and presents a much more temperature resistant environment as opposed to being shipped as a liquid. My dual chamber nasal spray is also being looked at for shipping vaccines worldwide to countries with tropical climates and where professional medical practitioners may not be available to administer vaccines intravenously. This allows millions of people at risk to administer their own vaccines at a fraction of the cost of typical intravenous vaccines that require refrigeration during transport.

I am certainly already a user of the intranasal, intravenous and oral pathways for various anti-aging enzymes and supplements, as well as doctor prescribed drugs.

Transdermal patches which also aim to be cost effective and easy to use alternatives. However transdermal patches may distribute enzymes and other products into the blood differently with each person depending on skin type, fat content, skin composition, etc. There is also no way of knowing how much of the supplement has entered into the body successfully, so users of transdermal patches need to be mindful of those possible variations.

Of course, many other delivery techniques exist as well such as electroporation, inhalers or iontophoresis which we will touch on later in this book.

Which pathway is best for you? All options above offer pros and cons. Do you have problems swallowing pills? Does a nasal spray cause excessive sneezing or inflamed nostrils? Do you prefer your anti-aging products in a liquid form that you can drink? Fact is that everyone is different and all pathways have benefits and disadvantages.

Ultimately my goal is to keep you posted with up to date information as much as possible via social media and / or new editions of this book.

Arthritis Res Ther. 2009; 11(2): R49

Imagine the cellular sack that holds all your cellular machinery such as the mitochondria, chromosomes, cytoplasm, nucleus, etc. What if we could make those sacks and place whatever we want inside them such as resveratrol, NAD, pterostilbene? Scientists can make these sacks and what's more interesting is that these sacks are attracted to each other with a static charge. The sacks are made from what is called a phospholipid layer and react when near another phospholipid layer.

Imagine a bowl of water with two drops of oil floating on the top. When those two drops of oil finally drift and touch each other, what do you think will happen? Will they bounce off each other and travel in opposite directions? Or will they appear to be attracted to each other and form a larger droplet of oil on the water's surface? The latter is true and it is similar physics inside the human body with phospholipid sacks known as liposomes.

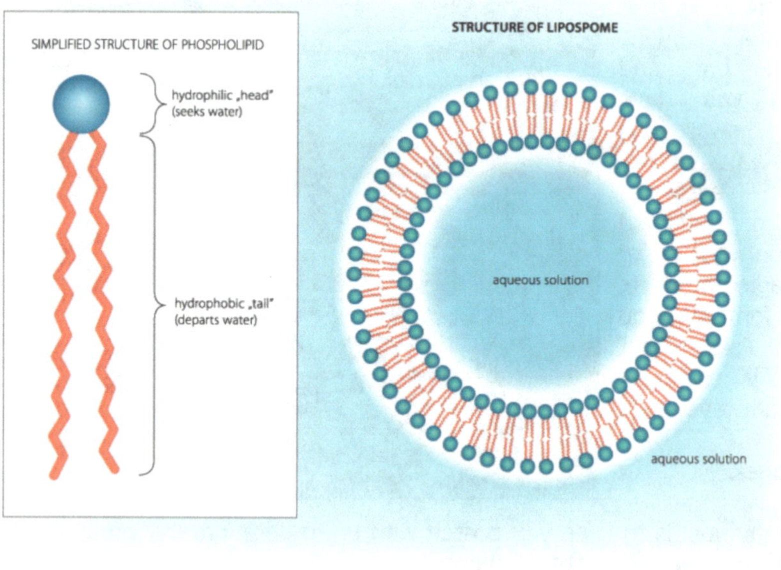

Scientists can put payloads of longevity supplements such as NAD or Glutathione inside these sacks for them to be administered orally. These sacks protect the fragile enzymes or supplements that you want delivered to your cells from being broken down inside the digestive system. This ensures greater absorbability into the blood stream to where the molecular payload should be delivered. This is incredible science as we can now deliver what we want where we want. I am (again) an avid user of this bio-delivery technology and will share with you what to look for and where to find these types of homing missile products later in this book.

Carnosine

Anti-Aging Score 3/10

Carnosine is already found throughout the body and in tissues such as the muscles, brain and kidneys. Numerous studies (in humans) have shown that carnosine deficiency is associated with various diseases including Alzheimer's disease, diabetes and cancers.

Carnosine is a great free radical scavenger which prevents cells from going senescent (zombie like) or from exhibiting dysfunction and creating inflammatory signatures in the surrounding medium. Because carnosine is such a great free radical scavenger, it protects the brain from oxidative stress, which is where its neuroprotection comes in (and why it is regarded as a preventer of disease).

Like most other wonderful products that our bodies produce to keep us young, carnosine levels drop with age.

- Senescent Cells – A zombie like state where the cell resists cell death and no longer functions properly, sending out improper signals causing a myriad of health problems.

Studies clearly show that when cells are approaching senescence that carnosine can reverse this state and cause them to act youthful and functional again. Carnosine can also prevent the formation of advanced glycation end products (AGEs), a subject that we have already covered earlier in this book, but which makes carnosine a great candidate for your anti-aging toolkit.

Vegetarians may have lower carnosine in their bodies, so I urge anyone eating a longevity diet as the one outlined in this book should supplement with this powerful amino acid.

Some final notes on carnosine are that it is able to reverse mitochondrial dysfunction; permits longer cell life to human fibroblast cells (skin cells) so that your face does not age as quickly, and it is also a great anti-cancer weapon. Studies have shown that carnosine can limit DNA damage, inhibit tumor growth and restrict the supply of ATP levels in cancer cells, resulting in the deprivation of the energy cancer cells require to flourish.

Nothing in this book claims to be a cure for cancer, however much of this book is a preventative measure for cancer and many other age related afflictions.

- Doctor's will try and cure you from disease, whereas your resident longevity experts will try and prevent you from getting disease in the first place.

We could harp on more about the wonders of carnosine in relation to blood pressure, bone health, physical performance, liver health, etc. But this book is to get you in and out as quickly as possible so that you are not overwhelmed with scientific studies or facts.

Carnosine is paramount for your anti-aging toolkit. What else is there to know?

The Microbiota aka Microbiome

The microbiome is the bacterium that resides inside our gut. Our knowledge base of this incredible machinery is still very immature. However we already know many things about the bacteria that flourish inside our bodies and there is no doubt that these little critters hold profound power over our longevity.

Note that the term microbiome originally meant the gene pool of the microbes in your gut (the microbiota) but these days microbiome is used loosely in exchange for microbiota.

Let's have a quick look at some of the science and how you can implement this data.

In 2019 a study published in The Journal of Physiology showed that specific types of gut bacteria were inflammatory in older mice (not young mice), and increased various pathologies but especially stiffness in the arteries. The offending bacteria was known as Proteobacteria (part of the Salmonella family) which created large amounts of an inflammatory product called trimethylamine N-oxide (TMAO) which inflicts damage to arterial walls and also causes oxidative stress. But inhibiting TMAO is easy according to a paper published by Chen & Zhang 2016 by The American Society of Microbiology. Chen and Zhang showed that resveratrol lowered TMAO significantly, a simple fix for a very silent and invisible cause of heart disease; yet another pathway to aging now defeated with sound science. Many other factors that cause high levels of TMAO such as high protein or high fat diets were also contributors to raising TMAO, but if you are paying attention to the diet advice in this book then we need not cover every angle of TMAO, except that red meat or high fat (animal fats) increases TMAO.

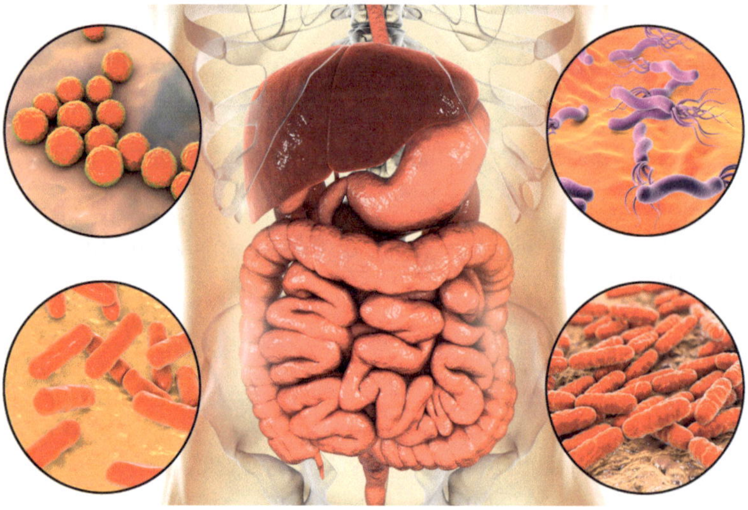

One thing we do know is that when we are young we have an extremely diverse range of gut bacteria. But as we age the types and the amounts of bacteria we have dwindle; science is showing that the loss of these bacteria species can influence early mortality (basically the more diverse your gut bacteria the better).

When the gut flora was examined in Chinese people over the age of 90 it was found that their bacteria were rich with bacteria known as Clositridium. The researchers of this study also found in other countries that groups of people of the same age (over 90) also showed increased levels of gut bacteria. The other countries studied were Italy and Japan. Even though the species of bacteria varied the amount of bacteria did not suggesting that older people with a weakened microbiome may be at higher risk of mortality. The variation in species is most likely a product of diet and environment. The less microbiota in your gut, the sooner you die (if we believe the evidence).

There are many byproducts and beneficial processes from gut flora such as metabolites or the creation of short chain fatty acids which aid in a myriad of healthy functions such as mineral absorption. So it is not beyond the realm of biology to think that some of these gut bacteria may assist or impede your health. Some of these little critters assist with breaking down food and even converting some foods into complex molecules that you otherwise could not get from diet alone. For example a broccoli compound (glucoraphanin) is converted into the powerful antioxidant sulforaphane by your microbiome which otherwise would be unusable when eaten. A wide spectrum course of antibiotics can wreak havoc on your microbiome and impede your ability to absorb nutrients, so antibiotics should only be used when absolutely necessary. Try and use antibiotics that target the harmful bacteria so that you are not wiping out all the good ones as well.

- Controlling specific species of gut flora is key to navigating longevity effectively. Ignore the effects of the microbiome at your peril.

It is too early to tell you exactly what strains to promote and what strains to fend off, however what we do know is that anything that decreases the number or species inside your gut is connected with early mortality. On the next page we will show you some strains to be mindful of.

One more thing - There is overwhelming evidence that the type of gut bacteria you hold controls your mood and personality. Yes, humans could simply be hosts for other lifeforms that control us. Our small intestine is also referred to as our second brain (hence why you get gut feelings when in a dangerous or stressful environment). We now know that Alzheimer's disease is connected with a type of bacteria that starts off inside the mouth and can move into the brain.

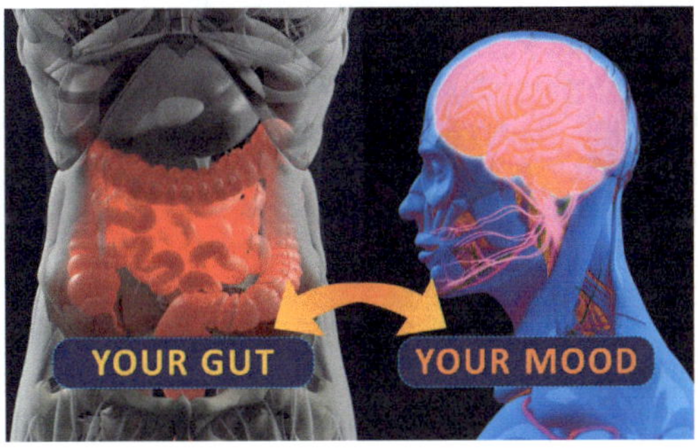

When mice were separated into two groups, risk taking versus docile, their gut bacteria was analyzed, then they were given antibiotics to kill their gut bacteria. Scientists then gave these mice the bacteria back, but in the opposite groups. The results were scientifically ominous; the mice exhibited the opposite behaviour. Risk takers became docile, docile became risk takers; a profound study that clearly demonstrates that gut bacteria can influence personality, including mood. How is this connected or relevant to longevity? Well if you are always in a poor mood and do not really know why, then apart from the cortisol levels and lack of endorphins silently killing you faster, it could be that your microbiome is poorly balanced from a poor diet or medications or something similar. Even depression is connected with gut bacteria and these types of things point to an earlier grave. Stress shortens your telomeres faster as well, so again, state of mind (tranquility) is important.

It is clearly established in the scientific literature that high fat, high protein or high sugar diets can impede the gut bacteria from working properly. Even poor sleep from jet lag can influence the gut bacteria and this is bidirectional, meaning poor gut bacteria can affect the quality of your sleep (Wang, Kuang 2017).

Other evidence shows that the development of a newborn's immune system may be hampered by cesarean section. The bacterium found in the birth canal plays a crucial role in early immune development (Knight 2017). So with that knowledge do not be afraid of letting your child get dirty or be exposed to the natural world (animals, other people, etc). It seems the evidence demonstrates quite clearly that exposure to bacterium assists with teaching the immune system how to defend against pathogens later in life. Even breast milk is laden with prebiotic carbohydrates to promote the growth of the Bifidobacterium species. This means our genetic code is written to ensure a vibrant microbiome from birth (our biology is reliant on the microbiome from birth).

Note - a prebiotic is a carbohydrate that cannot be broken down by the stomach alone, gut bacteria is required for prebiotics to be broken down. These prebiotics increase the amount of beneficial bacterial populations inside the gut and mainly come from (you guessed it) a vegetarian diet, legumes, whole grains, fruits, vegetables and nuts.

Parkinson's disease is also attributed to disturbances in the microbiome. Scientists took fecal matter samples from Parkinson's sufferers and transplanted that fecal matter into mice. Those mice began to show manifestations of the disease such as improper movement and ability. When the same team of scientists transplanted fecal matter from healthy humans in mice, the mice did not show these symptoms (Sampson 2016). These types of studies are quite scary in the sense that they point to colonies of bacterium being responsible for deleterious pathologies we see in old age.

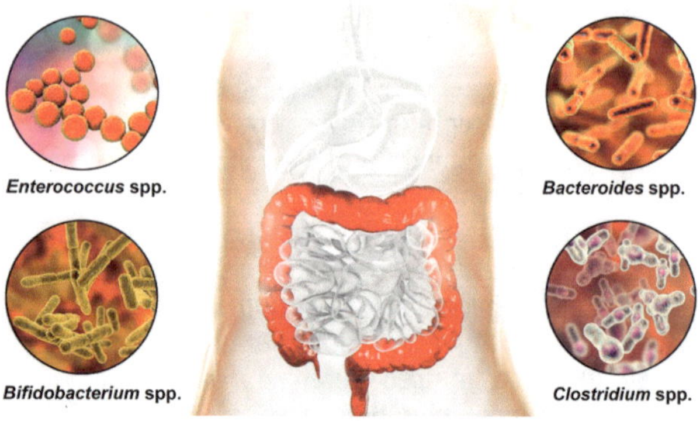

Note that fecal transplants (from one healthy person to a sick person) are now becoming standard as doctors attempt to balance people's microbiome by literally inserting fecal matter through an enema or tube. And (this will most likely gross you out), the same effective results can be achieved by placing that fecal matter into a capsule for people to swallow, ouch! (Kao 2017)

Exercise is also shown to increase gut diversity, whereas sedentary people seem to have less gut bacteria. This is important (if you have been paying attention) because we have already shown that there appears to be a connection in early mortality and lower levels of gut bacteria. The evidence is - the more diverse the microbiome the better for longevity.

Myself and a talented team of scientific cohorts are currently working on a novel method to control the good, the bad and the ugly parts of the microbiome. Stay tuned.

Ref: Chen M-L, Zhang Y, et al. mBio 7(2):e02210-15.
Ref: Knight R et al, Annual review of genomics and human genetics Aug 31 2017;18:65-86
Ref: Sampson TR et al, Cell. Dec 1 2016;167(6):1469-1480.e1412.
Ref: Kao D, et al JAMA. Nov 28 2017;318(20):1985-1993.
Ref: Wang Y, Kuang Z, et al (New York, N.Y.). Sep 1 2017;357(6354):912-916

A Quick Reflection On the Science So Far

Though very serious science has been done all over the world to make these discoveries about the compounds that can extend life, what is interesting, that most of everything we have looked at so far already occurs in nature.

- Metformin - from the plant Goat's Rue.
- Telomerase - from the Astragalus Root and also sold as Cycloastragenol.
- NAD - Already occurs inside every cell of our bodies.
- Spermidine – Green pepper, cauliflower, broccoli, mushrooms.
- Urolithin A - Pomegranates, nuts, some berries.
- Collagen – Fish, egg whites, berries, red and yellow vegetables.
- Vitamin K2 (Mk-7) – Natto, sauerkraut, hard cheers.
- Sirt 6 Activation – Activators found in olive oil, berries.
- AKG – Produced in the mitochondria.
- Creatine – Found in meat and fish.
- Curcumin - Derived from the Turmeric root.
- Hyaluronic Acid – Soy products, citrus fruits.
- NAC - seafood, chicken, eggs, broccoli, onions, and legumes.
- Pterostilbene - Almonds, grapes leaves and vines, blueberries.
- Resveratrol - Found in grape skin, blueberries, mulberries, raspberries, etc.
- Melatonin - Natural hormone which we produce for sleep.
- Quercetin - Found as a pigment in plant foods such as fruit and veg.
- PQQ - Found in kiwi fruits, breast milk, beans, celery, parsley, etc.
- AMPK – Found in every cell through the body.
- Ubiquinol aka CoQ10 or Co Enzyme 10.
- Astaxanthin – Found mostly in sea life such as algae, shrimp, salmon.
- L-Arginine – Nuts, seeds, legumes and chickpeas.
- Acetyl L Carnitine – Synthesized in the brain, liver and kidneys.
  Rapamycin – Found on the island of Rapanui.
- Fisetin – Found in fruits and vegetable.
- Exercise - Lengthens telomeres
- Diet - Eating non-processed non-animals based foods and keeping as close to nature as possible extends life.
- Fasting - Promotes anti-aging effects as opposed to gluttony which speeds up aging.

Everything so far appears to be around us or inside us already. It is almost like nature has mapped out everything and we just need to find them. This book is the culmination of the best scientific minds and research institutions across the globe, so it is quite interesting to me that after all the research and effort natural substances appear to be front and center. Because extreme longevity is seen as one of the final frontiers for humankind, so it's good to know that naturally occurring compounds are making a dent.

Note that the compounds you see online portraying to be the above list may not be the type you need, so select wisely and do your research.

And when it comes to purchasing these products, which are better, ones extracted from plants or ones that are synthesized in labs? The honest answer is whichever works best. There is no difference between a molecule found in nature and one made in a lab once they are floating around your bloodstream, they will have the same effect if they are the same molecule. Arguments can be made that molecules found in nature could come with pesticides, heavy metals, fungicides, etc, and similar arguments could be made regarding molecules made in labs. The manufacturing of either type is extremely well controlled for human consumption, and I am trained in the manufacturing and quality control of these products as well. Even though these molecules were found in nature, that does not necessarily mean they will work better than man made molecules in the lab, so do not have your lack of scientific knowledge exploited by those who pretend lab made molecules are dangerous or less effective. A molecule is a molecule, period.

And whilst it may be nice to learn that the list I've just shown is mostly from natural sources, I should note that to really extend life beyond the limits of your natural life, synthetic technologies will be used. From gene editing, cellular reprogramming to tissue reprogramming (which is what my company is working on).

To push past your body's limit of a natural lifetime is going to require something that I refer to as Tier 3 technologies. Oral supplements and lifestyle changes are great, but to really extend life, technologies that can reprogram the very makeup of human biology is required. These types of technologies are closer than you think as well. You will learn shortly about why I call these technologies Tier 3.

Those of you that resent the idea of using synthetic technologies will age and be more susceptible to disease, and those of you that embrace and implement the science will be very well positioned to remain youthful and vibrant.

Other challenges will exist with extreme longevity, we can only imagine what side effects a person that appears young but is 150 years old may have, and science is well positioned (especially by then) to figure out relatively quickly what is going wrong and how to fix it.

The power of science doubles each year, this is called Moore's Law. This bodes well for those of you that have a few decades left. You can expect to see Tier 3 technologies on the market well before 2030. And we are working with regulatory professionals to make sure our technology is available to the masses before the trial is even finished. This is possible based on the Right to Try law, where if you are sick and medications are not working, you can opt in and use technologies that are at the clinical trial stage to see if they can assist. These laws were first signed in by Donald Trump in 2018, and since, literally dozens and dozens of other first world countries have followed suit. No more will some politician get to dictate what you can use when you are sick or worse, on your death bed.

My technology is very safe, and I cannot wait to show you all. Maybe you picked this book up and my technology is already all over the internet, or maybe you just bought this book online and my technology is still under wraps. Either way, the future is bright, as other companies (not just mine) vie for market positions in one of the biggest biotech industries in the history of mankind (age reversal).

So What Is The Perfect Age?

It is interesting that the research on what people think is the perfect age has delivered all many different answers.

S.Jay Olshansky, a Professor in Epidemiology from the University of Illinois Chicago, studies this interesting area. If a pill was available to stop aging in its tracks, when you would you first start taking it? When Professor Olshansky asks his relatively young University students when they would start, those students think that 30 is old and would start taking it straight away (in their early 20's). But when Olshansky asked his 95 year old father the perfect age to start taking this wonder pill his answer was 50. His logic behind taking it at 50 is because the kids have grown up and Olshansky's father was still in good health.

63-year-old Professor Olshansky also thinks that 50 would be the perfect age because that was before he started having any age-related aches or pains.

Other researchers have also found interesting results. Many of those in their 50's do not want to be 30 again. The most satisfied or happiest age group (believe it or not) are people in their 70's. This is possibly because 70-year-olds are wiser, no longer need to stress about work, bills, relationships, etc. This research also shows that physical pastimes such as surfing or mountain climbing hold very little importance on happiness. It also shows that interest in hunting grounds such as night clubs and meeting your next partner hold very little importance as well. These things are in stark contrast to how a 20-year-old may think.

Predictably and coming as no surprise the research across all these age groups shows that no one wants to look or feel old.

So, is there a perfect age? According to Laura Carstensen, Director of the Stanford Center on Longevity, there is no perfect age. Laura says that there are too many variables and that it depends on what you think is important. You may think having large groups of friends is the greatest time to be alive, so you may lean toward your late 20's, or you may think that when life satisfaction (kids, wealth, wisdom) is greatest that that is the best time to be alive, therefore you may choose 55.

A combined study between Princeton and Stony Brook Universities and University College London found that worry and anxiety follow us around from the age of 20 to 50. From age 50 humans begin to experience an increased feeling of wellbeing right up to 70.

The trend is that the younger you are, the younger you will think the perfect age is, and alternatively the older (wiser) you are, the perfect age is older, around 50.

If we were to combine the data on this page, it is that people want to be fit and healthy <u>but not more than having wisdom, life experience and wellbeing</u>. So live life well, experience everything there is, good and bad, acquire wisdom, acquire enough wealth to be comfortable, and then of course, implement your anti-aging toolkit to slow down aging and remain 'physically' younger and healthier for longer.

Life Expectancy versus Geographical Regions

The geographical regions that offer the longest life spans are quite spread out across the globe which removes many environmental or climatic effects.

According to the United Nations World Population Prospects 2015, the average lifespan worldwide it is 70.5 years.

Monaco has the highest life expectancy with males living 85.6 years and females at 93.5 years. Japan is second with males at 81.7 and females at 88.5. And again ladies win in Singapore with males dying at 82.3 and females at 87.8. Women certainly win the life extension challenge. In our table below fourth place is Macau China, a gambling Mecca or as many know it, the Las Vegas of Asia. Fifth place is San Marino followed by Iceland. See for yourself -

Rank	Country	Average Life Span	Male Life Span	Female Life Span
1	Monaco	89.5	85.6	93.5
2	Japan	85	81.7	88.5
3	Singapore	85	82.3	87.8
4	Macau (China)	84.5	81.6	87.6
5	San Marino	83.3	80.7	86.1
6	Iceland	83	80.9	85.3
7	Hong Kong (China)	82.9	80.3	85.8
8	Andorra	82.8	80.6	85.1
9	Switzerland	82.6	80.3	85
10	Guernsey	82.5	79.9	85.4
11	Israel	82.4	80.6	84.4

This data is taken from the Central Intelligence Agency's World Factbook (2016).

Something which sticks out to me is that most Middle Eastern countries that practice fasting for a religious festival are not seen until we are into the forties rank. The only exception is Israel which comes in at number eleven, even though most Israeli citizens most likely do not practice fasting for religion. This is consistent with the science that fasting for anti-aging needs to be monthly or bimonthly or even weekly to have an anti-aging effect (once a year simply is not enough to deliver any results to your anti-aging toolkit).

I could derive many reasons about why those top eleven countries appear to deliver long lengthy life spans, but we will stick to science (and not sociology) and say cutting edge health care systems and lower cortisol levels from living in sophisticated societies may be the secret.

Testosterone and Mortality

Testosterone has been touted many times by wellness and longevity clinics to be a key component to staying young. And yes, you need a certain amount of testosterone to maintain muscle mass and also to retain a sex drive.

However, do not rush in and think testosterone is a magic bullet. Too much testosterone can end your life early and so can too little. Those that inject testosterone must be careful not to have too much. There is a magic number of testosterone to have for longevity which is between 300 nanograms to 1000 nanograms per deciliter of blood.

Now with that said, low levels of testosterone are also highly risky as well. There is a sweet spot to aim for, and we will take a quick look at this confusing science.

Eunuchs that have been castrated lived on average longer than their non-castrated testosterone producing counter parts by an average of 14 to 19 years according to a study published in the scientific journal Current Biology. The study looked at records from the 14th to the 20th century, and the evidence was quite clear on the life extending effects of early castration. So believe it or not there are many studies which show castrated men living longer.

Another study compared life spans of castrated men from an institution in Kansas for the mentally ill and showed an average life extending effect of 13.6 years, and note the younger you were castrated the longer you lived. However, I am not about to tell my male readers to reach for the closest pair of scissors just yet, as we can hopefully offer less painful longevity techniques than taking the big snip (we seriously hope).

Now in stark contrast to being castrated, low testosterone causes early death as well. Those men with extremely low levels of testosterone were prone to early mortality.

If Eunuchs live longer but the latest scientific research also shows that men with higher levels (in the top 25%, over 750 nanograms per deciliter) live longer, how can we possibly interpret this data? Both appear to work.

Well testosterone levels start to taper off after we are about twenty at a rate of about 1% per year. A study by Dr. Khaw, a professor at the University of Cambridge, Britain, found great advantages in males with naturally higher levels of this hormone. The study found that those in the top 25% had 41% less risk of heart attack, cardiovascular problems, cancer, and stroke when compared with males who had lower levels.

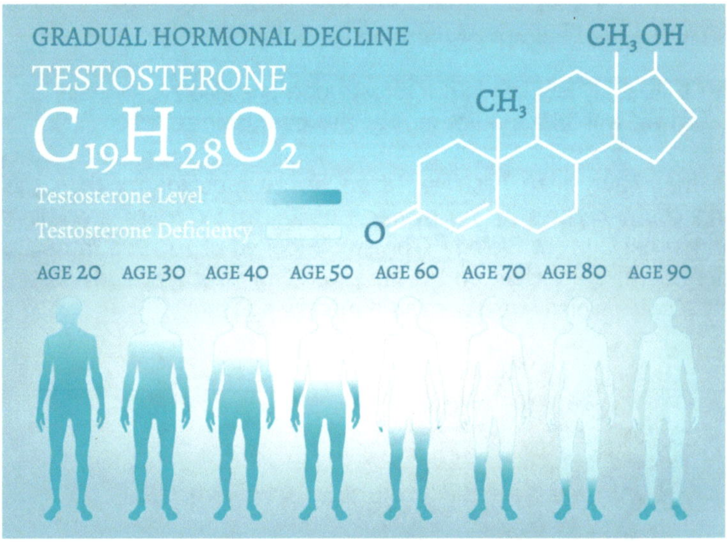

In fact Dr Khaw has even said that low testosterone levels could be used as a predictor of early death.

So how is this possible that Eunuchs can live longer if the evidence says higher levels are better? It seems either zero levels (Eunuchs) or robust levels 750 nanograms per deciliter work well, but low levels, between 50 and 300 nanograms per deciliter appears to be where the danger lies. Biology is full of idiosyncrasies like this that may not follow common logic (welcome to science, where we must boldly go where the data takes us).

Note that male pattern baldness is caused by dihydrotestosterone (DHT) which is a byproduct of testosterone. Baldness in women can also be caused by DHT. So next time you see someone's receding hair line chances are they have higher testosterone levels than their hairy friends.

J Endocr Soc. 2018 Apr 1; 2(4): 322–335.
Curr Sex Health Rep. Author manuscript; available in PMC 2015 Dec 1
Am J Cardiol. 2018 Jun 1;121(11):1321-1327. doi: 10.1016/j

Sex Drive and Lifespan

Now that we've talked about Eunuchs, let's talk about sex drive.

An interesting study published in the journal PLOS One infers that males with low libido are almost two times as likely to experience early mortality. This means, if you are male, and have no interest or a low interest in sex, then you double your chances of early death.

Scientists from Yamagata University observed more than 20,000 people, around 8,500 males, and around 12,500 females — in Japan for over ten years to study the links between sexual interest and "all-cause" mortality, which also included conditions such as cardiovascular disease and cancer.

Its common knowledge that sexual satisfaction is regarded as a benefit to psychological health and wellbeing in older groups, but the connection between sexual interest and longevity had not been directly researched.

The participants of this study were requested to complete a sexual interest survey during their regular check-ins. Subjects that exhibited low sexual interest tended to be older, consumed more alcohol, had diabetes, did not laugh as much, exhibited forms of psychological discomfort, and tended to have lower education levels.

Men in their forties and above with low libido were almost twice as likely to die from cancer, and there was even an increase of up to 1.36 for them to die from heart disease.

Interestingly, males with lower sex drives also reported lacking ikigai, which is a Japanese word which equates to a life worth living. These males exhibited a lack of ikigai. Much of the data in this study is already corroborated by what we see in other studies, such as loneliness is a fast killer. Strong family groups, or even strong social groups seem to help with longevity. A meta-analysis, called "association of loneliness with all-cause mortality" demonstrated that the effect of loneliness affected men's health more than females' health.

So, continuing with this Japanese study for low sex drive, researchers believe that males with lower sex drives, had those lower sex drives due to bad lifestyle choices. So does this means it could actually be the poor lifestyle contributing to early death and not a lack of sexual interest? Well in this study they did consider variables such as obesity and smoking.

And what we do know is that exercise and healthy living does boost testosterone levels, which raises a person's sex drive, so some of the lines here appear blurred and maybe things aren't as clear cut as we would like them to be.

What we do know is that healthy eating habits, exercise and socializing with family and friends are super important. So for you males out there, and I am especially talking to males here, if you are a bit of a hermit crab, go get that new lease on life today, and remember with social and family groups, its about adding positive energy, so steer clear of conversations around politics and other explosive topics, because you'll most likely give yourself more stress which isn't good.

Ref : PLoS One. 2022 Dec 14;17(12):e0277967. doi: 10.1371/journal.pone.0277967

Meet Dr Jeff Life

The good Doctor is a PhD and an M.D. But more importantly to you, Jeff was 81 in this photo. Does Jeff look old and frail? He actually looks incredible and this is achievable for any of you reading this book.

Anyone can maintain a youthful physiology if they really want to. Of course Doctor Life (his real name) trains much more than you need to in order to activate all the longevity pathways that training offers, so you do not have to spend one hour per day in the gym, however Jeff is the epitome of what is achievable now for those over 80. I think most 40 or 50 year olds would think that Jeff's physique is already out of their reach, but it certainly is not.

Dr Life's physique is actually keeping him young. Muscle requires energy to simply just exist, so he is also burning fat by holding extra muscle at this level. You, the reader do not need to hold this level of muscle as part of your anti-aging toolkit, some basic training (shown toward the end of this book) is all that is required to kick in various metabolic and longevity pathways to fight the aging process.

However showing you Dr Life's physique and age really hits home of where you could or should be.

Age related muscle loss usually begins after the age of 30. However performing some basic (and quick) exercises can ensure you move into middle and old age quite fit and strong. The fascinating thing about aging is that you get to completely decide how you age.

The less muscle you have, the less mobility you have, so the harder you will find it walking upstairs or doing other basic things. So that said, imprint this –

- The less muscle or the weaker you are, then quite possibly the lower quality of life you may experience.

Mobility is a serious contender when older people are asked about their quality of life. However mobility is malleable, improvable and adaptable.

Get your family elders into the garden or moving around more. I cannot stress the importance of keeping older people fit and strong.

Note that decreased mobility and strength also increases the chances of injury in later life. So again, this cascade of muscle loss affects quality of life in many other ways. Your best defense is to stay strong and healthy.

I met Dr Life at an anti-aging conference in Las Vegas. I managed to convince him to let us interview him for a television series that was canned due to Covid.

Get another eyeful of the Dr Life's physique before we move onto the next page. Nobody should ever say "I'm too old for exercise". After reading this page, you no longer really have any excuse to be in poor physical condition.

Many Babies Shorten a Woman's Life

Numerous studies have been conducted on childbirth and longevity and it certainly seems that the data points to the fact that females who have many children will also have shorter telomeres, thus reducing the amount of times their cells can divide and take them into old age. This data from Anna Pollack who is a researcher from the George Mason University in Virginia is one among many other studies that conclude similar data.

This is also consistent with our discussion two pages back regarding life expectancy in particular geographical regions. Those countries that have higher birth rates have much lower life expectancies (politically incorrect statement maybe but the comment is data driven and relevant to your anti-aging toolkit which is why I am including it). It is my job to show you all known pathways for aging and if you are a female currently reading this book that considers longevity important, then this information is going to be important to you.

Note that many things can shorten your telomeres (apart from having children) such as smoking, lack of exercise and just being overweight. And note, you can increase your telomeres with telomerase, meditation or exercise. However having five or more kids and you probably will not have the budget or the energy for the solutions to this problem (let alone time to think about it). And remember cortisol (your stress hormone) speeds up aging and your kids will ensure you have lots of cortisol to play with.

You're welcome!

Moving on -

Radical Life Extension Technologies

There are more radical life extension technologies that are already used and well known, however whether they will work is yet to be seen. Over the next few pages we will explore some of these technologies.

A quick look at the gene editing tool CRISPR (pronounced crisper). We will also cover some bio-hackers, such as Patient Zero and the Butterfly Experiment. We'll peer into the world of Fullerenes and Electroporation. We will also look at the world of cryogenic preservation and what that holds for us. The Yamanaka Factors, Cellular Divisional Limits and Biohacking is all covered in the coming pages.

This book would not be complete if we did not cover mTOR Inhibitors (aging pathways) Parabiosis and Apheresis which ensure you have the stem cells and other biological tools to offset aging. GDF-11, Nootropics are also worth a special mention. And the ultimate transcendence in life extension, mind uploading, going biological to digital. We will also take a quick look at what extreme longevity will mean for humanity and the Earth.

These radical life extension techniques may sound like science fiction movies but to those in the know, they are either already here or coming very soon.

So read on.

Biological Escape Velocity – Immortality Escape Velocity

You may hear anti-aging titans use the term "immortality escape velocity" or "longevity escape velocity". The word immortality should be regarded as metaphorical or symbolic, but never literal. And you probably will not ever see immortality achieved with just a simple pill. It will be a range of treatments that remove cellular damage and control gene expression among many other facets. Instead of immortality being some type of pill that magically stops us from ever growing old, it is more likely that science will invent numerous treatments to outsmart and prevent disease in the first place. Once we can outsmart our biology enough so that we do not even get disease and can keep our cells young and dividing (the latter of which it seems we've already done), we can keep ahead of the things that will kill us, and that is known as 'immortality escape velocity'.

However, I modified this term slightly to **"biological escape velocity"**. For me, we are trying to outrun biological decline, we are trying to escape from biology as it stands, not escape from longevity or immortality.

But is immortality possible? Well immortality or biological escape velocity is closer than you think. All those science fiction movies set 100 years into the future where people grow old are,,, well,,, wrong! That reality will not be happening as you'll see in coming decades. Hollywood pay attention! Remember we have already doubled human lifespan in the last 120 years (with very basic technology), there is no reason to think with our new super sophisticated science that we are not overdue for an explosion in human longevity.

If aging is gravity, and anti-aging technology is the propulsion, then escape velocity is just a matter of overpowering gravity (the causes of aging).

$V_e = \sqrt{2GM/R}$

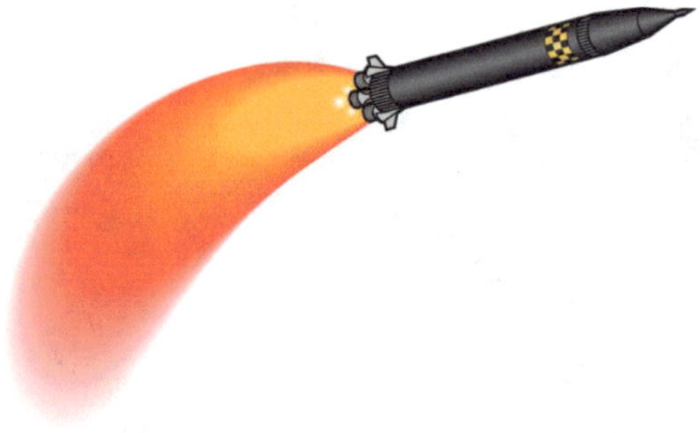

Some of the brightest minds in this field are predicting that we will reach escape velocity (V_e) before the year 2030. Frustrating though, this is one field of science where we could achieve V_e but it will not be proved until people are living to 150 or 200 years old, because how do you prove something works now, when it takes 100 years to demonstrate?

So V_e may even be here already, but no one can say it until current trials see people living well past 100 years with youthful function.

The Ray Palmer Scale For The Intervention on Aging

Where do you currently sit on my anti-aging scale? Types fives (those that are moving through life without any particular protection to dying) down to Type ones (those that will not die).

Type V The type five category do basic things to stay alive, they wash their hands, wear helmets on a bike, seatbelts in cars, their diet is fast food and occasional healthy food, but overall they do very little to extend their life or health span.

Type IV This demographic exercises, eats relatively healthy. They try and steer clear of toxins such as cigarettes, maybe limit alcohol but do not practice any serious anti-aging knowledge.

Type III This demographic is proactive in using anti-aging technologies. They study books like this, they use enzymes and foods specifically known for their anti-aging properties, they possibly practice fasting, most likely exercise, have a meticulous diet and steer clear of any biological insults.

Type II This group ingests, injects and uses fringe technologies such as gene therapies, modified viruses, Bucky Balls, GDF-11, exosomes etc. They use invasive technologies that are still unproven and possibly very dangerous but still may hold profound anti-aging effects.

- Note, the moment these Type II technologies are deemed safe, anyone using them becomes a Type III. The Type II category is risk taking technology. It is that scientific grey area where humans push the limits of what is and is not safe.

Type I The ultimate state of biology. Type ones have achieved biological escape velocity and can stay ahead of whatever biological decay may come their way. Could Type I's using Type II technology already exist without even realizing? Could we see this category populated in the next few decades? It's not only possible but may be probable.

Current technology allows you to get safely onto the Type III and not so safely onto the Type II level.

Type I is the Holy Grail. I am a seasoned Type III, and by the time you read this, I'll probably be a Type II.

You can read my paper on this aging system at the following link –

www.pubmed.ncbi.nlm.nih.gov/35309156/

Just hit the View PDF button for the full text.

CRISPR-CAS9 (Type II Treatment)

Think of molecular scissors taking snippets of DNA out of your genome so that you no longer carry those disease or aging genes that are built into you to program your death. This technology has exploded for all manner of disease control right now, so get used to that name CRISPR. If you have not heard of it, you will.

Using a carefully engineered protein called CAS9, it becomes possible to edit and cut out any gene or cluster of genes by programming CAS9 with a section of the DNA you want to remove. We are basically using techniques that have taken evolution billions of years to master. This process has been happening in nature since the dawn of time. It is how bacteria have been fending off viruses. We've simply taken one of nature's incredible tools to become the masters of our own genome.

CRISPR-CAS9 can edit genes with extreme precision. Now it just comes down to which genes or gene clusters we need to remove to pause aging. Those genes are quite well known at the moment, but CRISPR is still in the early stages and may be expensive. In the future however (maybe only a decade) this therapy should be at a level where it is cost effective in most modern countries.

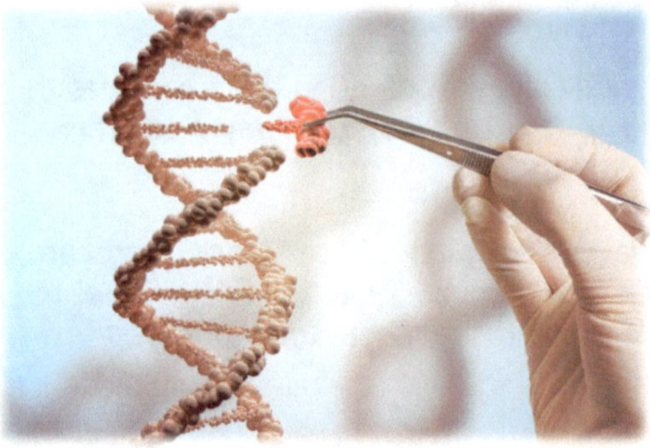

Note that CRISPR is tailor made for each individual. That is, a sequence of your genetic code is taken then a section of code selected and CAS9 gets to work to remove it.

CRISPR will become the norm for unborn babies as well, because ethically the argument will be made that if a child can be born without disease, why would we condemn that child to suffer by not implementing CRISPR on their genome? Disease control is the most obvious form of longevity, just like health care is a form of longevity. Denying any form of disease removal (health care) in a young child and allowing that child to suffer would be considered barbaric in coming years.

With CRISPR able to make humans immune to most genetic diseases and also able to remove the aging genes, extreme longevity seems very possible. Whilst we are editing your genome, why don't we make you immune to junk food and give you a faster metabolism to process the modern diet simultaneously? The human genome will be edited significantly in the coming decades by CRISPR and those changes will be passed down to the offspring, permanently changing who we are as a species.

If you think that's tampering with nature, so is antibiotics, vaccines, surgery, chemotherapy, Botox, among others. When there is a disease we've always tried to fight it. The ultimate frontier for human health is extreme longevity, but more importantly increasing the amount of years you are young and healthy (your youth span and health span) is what we all really want.

Note there are many other types of CRISPR now coming to the fray.

Evolutionary Escape Velocity - The Death of Evolution

Natural Human Evolution is dead. So, what does this mean? Well as we just saw with CRISPR, we now have various tools at our disposal to interfere with evolution. Never again will our species have to wait thousands of years for a random gene mutation to come along and either make us weaker or stronger. We can now edit genes at will and take vast control over our biology. Much of the world is still oblivious to this field of science, but what it means for you is huge.

For example, genes that regenerate tissues can be copied and put into you, or "knocked in", so that you may have two copies expressing a rejuvenation protein instead of one. And genes that make you old can be turned off or cut out, or as we say "knocked out". Gene editing is going to be huge in longevity and don't be surprised when clinics start opening up and offering anti-aging and life extending services to rewrite your genomic instruction book that literally gives you a longer healthier life. Just like Botox and skin lasering clinics or even test tube babies back in the 1980's was the stuff of science fiction at one time, this gene editing technology is coming to a clinic near you, like it or not.

There are two types of cells that can be edited. Somatic cells, which make up your tissues, and germline cells, those are the ones passed down to your offspring. Editing somatic cells still comes with risk and controversy, but editing germline cells, where we change the species slightly is a tinder box for science and everyone has an opinion on this.

Evolution on Earth has devised its own way to ensure life forms evolve and are adaptable to survive in their environment with genomic mutations or what's commonly known as natural selection. Through genetic mutations genes may give lifeforms stronger or weaker survival traits. Those with the mutations that see them become weaker may not survive long enough to pass on their inferior genome, if they do pass along those inferior genes, then we have the makings for a genetic disease. However, the mutations that cause a species to become stronger, increase the chances of survival, thus increasing the chances of passing down that superior genome to its offspring, and this is known as natural selection. The human race is now devising ways to ensure that inferior genomes no longer cause early mortality and that those inferior genomes are already being edited out with technologies such as CRISPR, base editing, PASTE, etc .

I have been saying it for many years, that we have reached an era that I personally call, Evolutionary Escape Velocity. I created this phrase because I think it captures where we are scientifically and our increasing power over nature. We have certainly escaped the clutches of natural selection and we will only get farther away from natural selection as time goes on. Farmers commonly cross breed plants and animals to make them stronger and healthier, and this is a form of evolutionary escape velocity, because we are no longer waiting for natural selection to make the changes for us.

Thus, no matter what your genome, superior or inferior, the death of the natural evolutionary pathway is already here. Humans will now control how their species will evolve with genetic editing and no longer will natural selection decide for us. Unborn babies that exhibit disease causing genes may now be genetically edited, so that no longer are they born weak with little chance of survival or passing on their disease-causing genes if they grow up. If their germline cells have been edited, then when those edited babies grow up and reproduce, the edited genome then becomes part of our new DNA, at which point our species is no longer 100% natural and we (as opposed to nature) then become the master engineers of our own genetic evolution. Carefully engineered human hybrids, mutants, abominations, call them what you will, but genetically stronger and healthier humans are now only a few years away, because the technology is already here and available for those of you that search it out and can afford it. Currently the cost is the only barrier, but this will fall rapidly with gene editing tools falling in cost.

Of course, the argument for gene therapies goes both ways. Should an unborn baby or a child be left to suffer with some heinous genetic disease? Or if a cure exists, is it unethical to leave that child to suffer and are we compelled to cure them if a treatment exists?

What healthy person gets to decide about someone else's suffering?

Natural selection is now redundant as evolutionary escape velocity now takes hold. Natural selection simply takes too long for any benefit during our limited life span, so enjoy being part of an era in human history that we have little choice in actually being part of; because for the first time in the known universe a species has become its own genetic master; we now choose what genes are passed on and what genes are not.

How does this help you? If you own genes that are faulty, it is relatively easy to find a CRISPR lab via online searches to ensure your survival; you finally have options, so get your genes tested. And if you struggle with that, feel free to reach out because we have a vast network of scientists that we could possibly put you in touch with, and my goal in life is to prevent suffering and disease, as opposed to just managing it.

Natural selection has changed forever, we have reached the point where natural human evolution is dead.

Okay, but seriously, so where is all this taking us?

So far we have read about all manner of molecules, diet and fasting, big deal.

We have just heard about evolutionary escape velocity and CRISPR, big deal.

But seriously, is there something more profound?

Is it possible with today's technology to reverse aging?

Is it possible that a 40 year old's bio markers
could become the same as a 20 year olds biomarkers?

Where are these fantastic types of technologies?

How far off into the future are they?

Is this just the stuff of science fiction movies?

Introducing our first biohacker, Patient Zero -

Liz Parrish - Patient Zero - (Type II) – Our First Biohacker

Liz Parrish is the CEO of an American company that specializes in cellular aging. What makes this lady stand out from the crowd is that she has taken a leap of faith and put her money (life) where her mouth is. Liz has become the first person to undergo two treatments to prevent aging. Those treatments are a myostatin inhibitor and telomerase gene therapy. Myostatin inhibitors simply prevent age related muscle loss and the telomerase therapy ensures that her cells can continue to divide and function well into old age (and maybe indefinitely).

Telomerase (that we have discussed earlier in this book) not only prevents telomeres from becoming shorter, but the resulting longer telomeres tell her cells to act more youthful (its more complicated than that really but for the purposes of this book it is sufficient). The concern has been that cancer cells may become upregulated by using telomerase, however by keeping your telomeres long you also keep your immune system functioning at a pristine level, possibly as it functioned when you were young, which is the best weapon against diseases like cancer.

It is worth noting that so far any correlation between telomerase ingestion or injection and cancer remains unproven. In fact so far the science has shown that cancer and telomerase induced longevity never meet (I can never give you absolutes; there is always room for new results, so caution advised). Liz is doing exactly what I am doing, and that is approaching longevity as a multi-pronged approach by combining more than one technology simultaneously. Patient Zero is obviously using technology that is considered fringe (Type II), putting her safety on the line, however Patient Zero is very vocal in accepting any outcome of such fringe anti-aging technology if it pushes the science forward and results in saving lives, so kudos and great respect for this brave Type II lady.

The reason that muscles lose mass when we do not use them, is from an evolutionary stand point. Over the course of millions of years, we evolved to lose muscle mass if we weren't using them, and this assisted with survival, as food was scarce and muscles have huge energy requirements, and its why body builders need to hit the gym daily to stay big. The myostatin inhibitor that Liz Parrish took allows muscle mass to remain longer, which will hopefully prevent Liz from becoming weak or frail later in life.

So what did Patient Zero actually do?

Liz flew from the U.S to Colombia where it is legal to perform this procedure where she received an injection into her muscles containing a gene known as follistatin, which is known to increase muscles mass. She also received a second injection of genetically modified viruses that would lengthen her telomeres. These types of treatments are not for the faint hearted (especially when proper human trials have not been conducted yet), so the reader needs to be sure the technology works. This procedure (according to Liz) took place September 2015, fast forward 6 months later, and Liz's telomeres are (allegedly) 9% longer, or in other terms, her white blood cells now function as if they were considerably younger. Will this have an effect on fighting disease? Probably, but only time will tell. As we delve into Type II's, I must offer this following information, that scientifically and medically speaking, the above results mean very little (if anything), as the experiment was only done on one person, not under professional observing conditions and it is still too early to know if there are any long term side effects. There is also a conflict of interest as Patient Zero is CEO of an anti-aging company promoting these technologies, however if the results are to be taken as worthy, then Patient Zero has achieved Type II status. The future is incredibly promising.

Fullerenes and Electroporation – The Butterfly Experiment (Type II)

Following in the footsteps of Patient Zero (well actually not following, but simply after), another pioneering biohacker or maverick (depending on your position) who is an accomplished scientist injected himself with another therapy which was completely untested on humans and certainly not permitted by the FDA. However this process was different. Electroporation was used in which two electrodes are placed at the site of entry and then an electric current is passed through the skin and tissue. This opens up micro-pores in the cell membrane allowing nearby molecules (that you may have injected in the surrounding area) to pass into the cell. This then (depending on what molecules you have selected to inject) can control the cell and gene expression resulting in a permanent effect on your physiology. The advantage this technique is that no virus is required to deliver the payload and the molecules you are injecting do not modify your genome. In this instance a plasmid was injected into the area to promote the production of human growth hormone (HGH) which declines with age to hopefully give this scientist more youthful levels of HGH.

Dr George Church of Harvard Medical School (very well known in Genomics) is now studying and monitoring what effects this therapy has had and will have on this pioneer.

Interested in this therapy and patient zero's therapy? Keep your eye on my social media account, I am studying the science with very talented great people.

Fullerenes (Buckyballs) are nanoscale (extremely small) carbon cages that make great platforms for delivering drugs into the body.

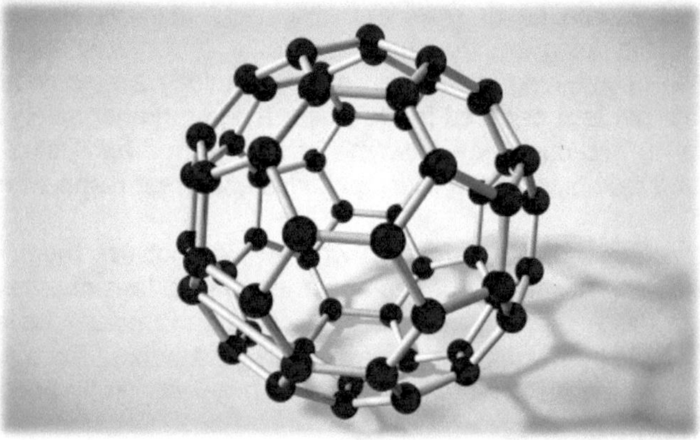
Fullerene C60 (60 carbon atoms)

Because fullerenes can be made at varying sizes (the amount of atoms), they can also be tissue selective on where they end up delivering their payload which could be genetically modified genes, viruses, antioxidants and more. This is nanotechnology meets biotech meets genetic engineering meets hormone therapy meets anti-aging, which is an incredible cross section of sciences and human engineering to pacify human nature's quest to live longer.

In a study by Dr Fathi Moussa of the University of Paris, rats given C60 Buckyballs lived twice as long as their natural counterparts!!! The science of why these rats (now known as Bucky Rats) lived twice as long is unknown and is only speculative, but like every challenge in science, there is also an answer to find in science. Could it become a Type I treatment? This experiment needs to be repeated and peer reviewed by longevity scientists to have the credibility and accolades it may deserve.

GDF-11 aka Growth Differentiating Factor 11 (Type II)

This book would not be complete without the mention of one more rogue treatment worthy of sharing where people are currently injecting themselves with this Type II treatment. Harvard scientists have managed to reverse aging in mice using GDF-11 which was published in the magazine Science 2015.

GDF-11 is a natural peptide that seems to be able to restore aging muscles, heart tissue and brain cells among other things. So understandably you can see why this peptide may become quite sought after.

At the time of writing, about 20 volunteers are undergoing GDF-11 treatment. GDF-11 is endogenous, which simply means that it is a naturally occurring peptide that the human body already makes. This also means that GDF-11 may be safer than using viruses and other technologies that are foreign to the human body. However in larger doses the science is still not proven to be 100%, but neither is aging 100% safe, so which path do you take?

This GDF-11 treatment appears to replicate parabiosis (page 85) by rejuvenating organs and tissue and it also appears to repair DNA damage which restores the stem cell back to a youthful state.

Note that the data shows that the effects of GDF-11 in young people (let's say 20 year olds) may be extremely small, however the effects on 60 or 80 years olds is profound. This data may simply show that young people already have sufficient amounts of GDF-11 to function youthfully.

To sober you up though, do not rush out and start purchasing vials of GDF-11, they are unproven, and even though you may feel great in the first year, who knows, after a year your eyesight may disappear, you may have the early onset of Alzheimer's from this treatment, or you could simply have organ failure and die. These Type II therapies should only be used under extremely strict medical supervision. However I hear your calls and will report via social media and tell you when and where to buy at the early possible moment.

The GDF-11 molecule is the largest signaling molecule inside the human body. It is not easy to synthesize so be extremely careful if you ever see someone offering to sell it. 50 nanograms (50 billionths of a gram) has been shown to causes nasty side effects such as mouth sores, diarrhea, shortness of breath and skin inflammation among others. This is not something you can just inject and be young; it takes serious science to administer properly. Point one (.1) nanogram twice a week is the amount used in the current trial. GDF-11 is Valde Fortis (Latin for very strong, so beware).

Cryonic Preservation

Cryogenic freezing is taking place all over the world right now. Freeze yourself before a disease kills you (cryothanasia) or right after a disease has killed you (cryogenics). Your blood is replaced with anti-freeze and you are snap frozen with the hope that in the future you will be woken up with new technologies. The snap freezing overcomes the problem of water expansion destroying your cells as well. You may think there is a very slim chance this technology will work, but a slim chance is better than no chance at all, so cryonic preservation is now a booming industry.

However, we already have proof of concept; there are multiple animals that can freeze and come back to life years later. The Wood Frog, Artic Wooly Bear Caterpillar, Darling Beetle and others have all developed ways to reanimate after being frozen. There are numerous ways these organisms achieve this, but the point here is that eye tissue, muscles, brains cells and other essential machinery remains functional after being frozen and then thawed. This is great news for cryonic sleepers.

Sleep is like time travel. You have no concept of time, you shut your eyes at night and what feels like an instant is 8 hours. Cryogenic freezing will be no different. If it takes a century or a million years to perfect the science, the concept remains the same; you will close your eyes and enter cryogenic sleep, and what feels like a split second later, you will wake up in a strange new world hundreds or thousands of years later that none of us today could hope to imagine.

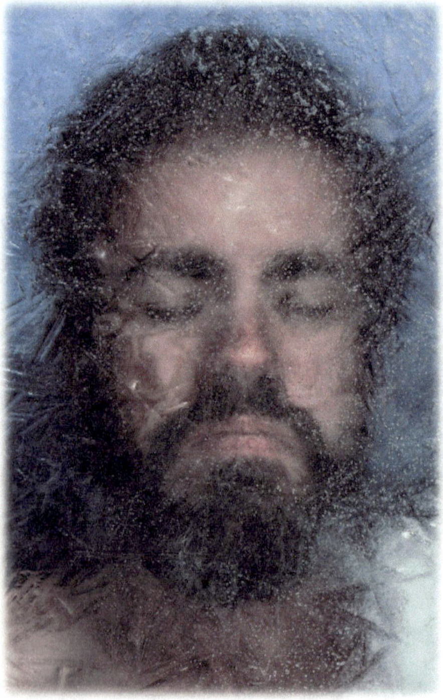

But what if you did wake up in a world hundreds or thousands of years later? You would be a freak show, grappling to come to terms with how different the world is. What if we woke up a Viking today, how would that Viking react? How would that Viking fit into our society? Humans today are a little bit more sophisticated than a Viking thankfully, and humans that are cryogenically preserved have the added benefit of knowing that the future will be vastly different, so waking up a century or two from now should not be so tragically shocking as it would be for our pilfering Viking.

Subscribers of cryogenics are abundantly aware that freezing the entire body could be a waste of resources, so just freezing the head where your personality and consciousness resides is (believe it or not) an extremely viable option for this community. Artificial limbs are already being built, so in one hundred or five hundred years who knows where things will be with artificial bodies. So cryogenic patients may not waste time freezing their entire bodies. This makes sense as it requires smaller resources which reduces the chances of system failure during the long sleep and also reduces the amount of biological material that needs to be revived (not to mention reduces the cost). It's an uncomfortable idea but it's well accepted in the cryogenic community.

Maybe something even more discomforting than human decapitation is that scientists at Yale University have managed to restore the circulation to severed pig heads, meaning a severed head could actually be kept alive. Science already knows how to oxygenate blood, pump it and perform dialysis on that blood etcetera, so the technology to pumping blood into a severed head to maintain its consciousness is in fact already here. It is just a matter of who and when will be the first human. There are obvious ethical issues of pumping blood into a pigs severed head. If the pigs were conscious, how do we know they were not feeling extreme pain or stress from there recent procedure? Professor Nenad Sestan of Yale University experimented on over 100 pig heads and is believed to have kept them alive for up to 36 hours. However brain scans showed no brain activity even though brain cells were kept alive.

Can you imagine in 20 years' time seeing an advertisement for a machine called something like a "Consciousness Continuance Incubator" where instead of buying a boat or holiday house, you spend $50,000 on this machine and as long as nutrient oxygenated blood is loaded into it each day your severed head can sit quite comfortably watching your favourite shows and surfing the web with only the movement of the eye. We are getting a little off topic but with the above research on severed heads it no longer seems so strange.

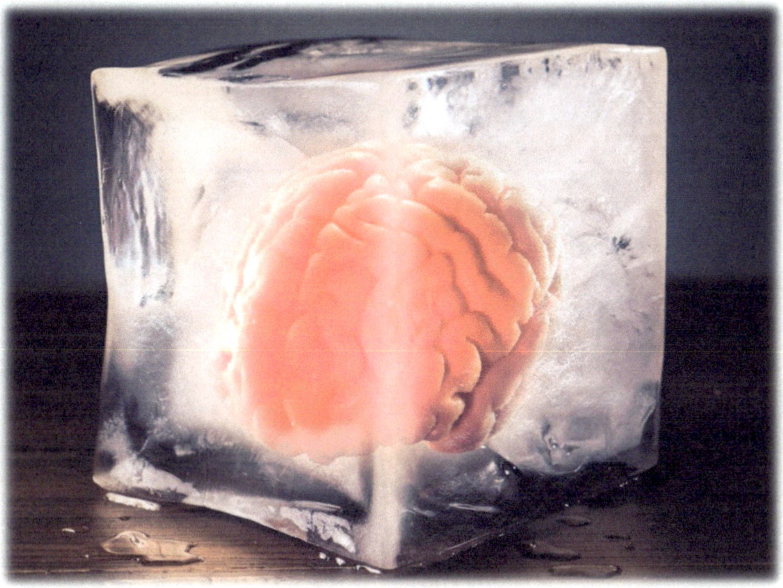

A colleague and one of the best known anti-aging experts in the world has already organized to have his head frozen based on the future being able to grow human bodies that a thawed head can be attached to.

This same expert also believes that people born today could live to 1000. So, it is not just me who holds hope. I chat with these scientists at anti-aging conferences and discuss many fascinating facets of emerging longevity technologies.

Many of you might think it is crazy to think someone can live to 1000 years, but I used to hear other professional astronomers say that we will never see planets orbiting other stars because they are too faint, which I always felt was a narrow view. It caused me to coin the following phrase when explaining science to others -

"Never predict the future with the limits of today's technology."

And guess what, within a few years later, we directly observed planets orbiting other stars (HD95086b)

Parabiosis / Aspheresis

Imagine taking blood and stem cells into your own circulatory system from a much younger person and this process rejuvenated you into a much younger version of yourself? Science fiction? Not so, it is happening right now.

Animal trials show overwhelming evidence that using the blood (and its components) from much younger animals of the same species has a profound effect on bone, skin, hair, heart function, muscles, brain and athletic activity (to name a few) in older subjects. Parabiosis has been around since 1866, so it's not new, but recent research shows profound effects in animals. There are numerous ways to do this effectively and some scientists are even patenting their techniques now as the awareness for this rejuvenating technology builds.

Some say that this technique is not reversing aging, but repairing damaged organs and cells to a more youthful state which I think is pretty much the same thing and the intended effect no matter how it is worded. The fact that it works to an easily detectable degree is extremely exciting for those that want to live stronger and longer.

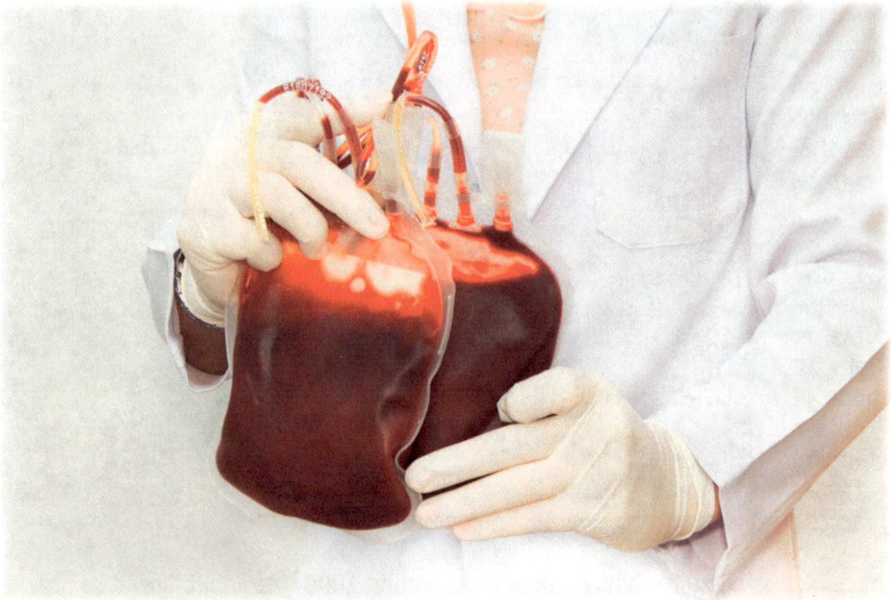

You may need a bit of blood though; "one pint of claret will not cut the mustard". There is every chance that you would require numbers in the realm of 2 litres per month for one year for parabiosis to have a profound effect, but it could extend your quality of life and lifespan by decades. With many young people happy to sell their healthy blood, you can expect this to be a mainstream option for those that can afford it.

Young Blood Plasma

There are many clinics that have opened up with the sole intent on injecting you with the plasma from young blood. Similar to parabiosis this technique only uses the plasma part of the blood. But does this have a bearing on rejuvenation? According to the recipients, you bet it does. Injections of young blood plasma appear to rejuvenate tissues and allow recipients to feel younger and more youthful. So does this mean that we do not need the entire blood package as in parabiosis? Possibly, but I cannot find any trials where the two techniques have been compared side by side under proper observing conditions.

Blood plasma is yellow in color and is what suspends the blood cells inside the blood.

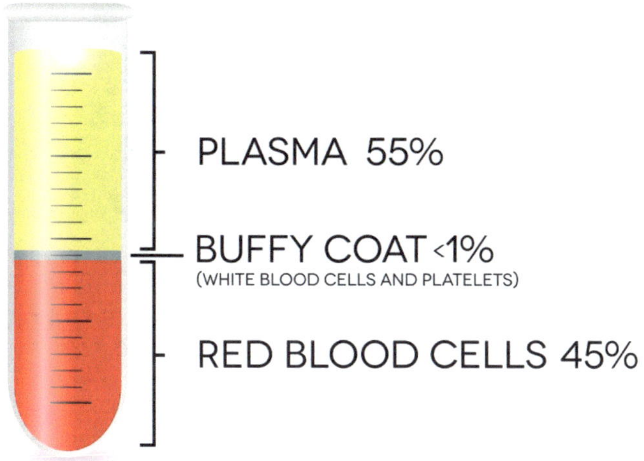

Clinical trials still need to be done to extrapolate exactly what young plasma can actually do, but because clinical trials usually require a placebo group as well, it is hard to fund something where fifty or more people are getting infusions of a liquid which is absolutely useless (the control or placebo group); it is expensive and dangerous, so clinical trials will be few and far between for such research.

There are certainly many claims from recipients that they feel better and more energetic, have better skin, breathe better, exercise better, etc, however professional clinical trials are needed to substantiate the claims. There have also been some question marks recently from Governments.

Like many things in aging, plasma starts to drop off in old age, so with the young population producing much larger amounts of healthy plasma proteins than older people, it was only a matter of time before this became a target of aging and a possible treatment.

There can be side effects to this treatment as well, but would I receive young plasma? Yes, I would (and of course I'll report on my findings).

Cellular Divisional Limits

The Hayflick Limit is the stop sign at the end of the cellular road that prohibits your cells from dividing further. However, this limit is not only appearing flexible but reversible with the Yamanaka Factors, parabiosis, telomerase and other technologies.

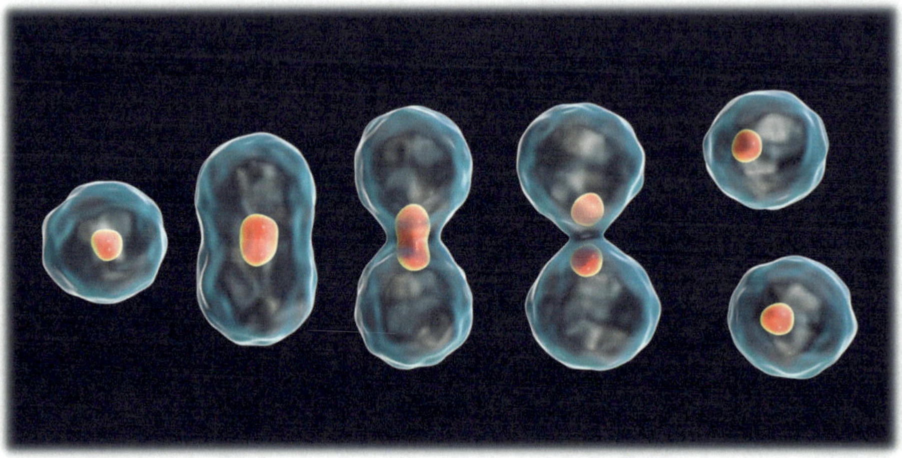

Recent biology was that living past 115 was impossible because we simply ran out of cells to divide to keep us young and healthy, however that argument is no longer true, not only can we reprogram existing cells to a youthful state but we can even take cells from young donors (parabiosis) to ensure we never reach the end of our cellular limits again. We can modify our genes, we can purge dysfunctional cells and much more.

The term "Hayflick Limit" was named after its founder, Leonard Hayflick in 1961. Hayflick had earlier refuted the popular belief by Nobel Laureate Alexis Carrel that normal cells were immortal; instead Hayflick demonstrated that human fetal cells would divide between forty and sixty times before entering a state of dormant unresponsiveness (senescence).

Hayflick's discovery led to new science and ways of thinking. Prior to this discovery the scientific community had believed that cells only stopped dividing based on their harsh environments which may not be producing the essential things for a cell's ability to divide. However, once this belief was smashed, it paved the way for technology decades later to be included inside your anti-aging toolkit.

Telomerase was then discovered (which we have already discussed), however other compounds and technologies are now in the mix to lengthen your telomeres which you will discover shortly.

The Epigenetic Clock

Biostatistician Steve Horvath has come up with a rather interesting tool to work out your exact biological age. The tool was originally devised by Hannum et al, which could determine the age of humans by looking at genetic markers (methylation markers). Originally known as the Hannum Clock, Steve has honed this technology and it is also now commonly referred to as The Horvath Clock. This clock uses epigenetics. The epigenome is the machinery that controls which genes to turn on or off. Using hundreds of epigenetic markers, Horvath can extrapolate from the methylation markers and determine an accurate age of your biology. This means that not only can your biological age (not chronological age) be determined, but that the Horvath clock can also tell you what you are likely to die from, not from genetics alone, but what markers in epigenetics are far enough advanced to indicate a particular disease is impending.

Because our bodies age at different rates naturally, epigenetics allow us to see into the future and know what organs are likely to fail first, years in advance.

Doctor Horvath feels that the exciting facet of this epigenetic clock shows that most tissues age at the same rate as expected, however two tissues do not age at the same rate as other tissues. The heart tissue seems to age differently and even female breast tissue appears to age faster than surrounding tissues.

The Horvath clock is always being improved, but to date it can say with more and more accuracy when your time is actually going to be up. This gives us advanced forewarning of any impending disease or condition and allows us to make radical changes to our lifestyles or environment to ward off any time limits we may have been sentenced to.

Epigenetics is not actually part of the genome or DNA but rather the control mechanisms over those genes, which in turn changes the cellular environment which then tell cells how to behave, so the genome is never changed but the gene expression (how those genes tell the surrounding cells to behave) is changed.

The importance of epigenetics cannot be underestimated, even Curcumin can instruct tumor suppressing genes to be activated to ward of cancer, and this is done through epigenetic mechanisms which may also be known as DNA methylation (among other terms).

You can read one of my papers on Aging Clocks and Mortality Timers at the following link –

www.pubmed.ncbi.nlm.nih.gov/35783114/

Just hit the View PDF button for the full text.

Epigenetic Diet – Epigenetic Drift

Anti-Aging Score 8/10

You have already heard terms such as DNA Methylation and Epigenetics. Well the epigenetic diet is basically food groups that are known to influence how genes express themselves in your body. Some genes can cause disease, others can prevent disease. The epigenetic diet is comprised of foods that are shown to activate longevity genes while at the same time downregulating disease causing genes.

The University of Uppsala, Sweden has shown that tea has been shown to upregulate genes (in women more than men). Scientists believe the results may simply be because women drink more tea than men, however the epigenetic changes were found in genes that influence cancer and estrogen production which also could determine the difference between the sexes. There are plenty of other health benefits from tea, so enjoy a cup when you can. However the same Swedish study found that coffee unfortunately does not appear to influence epigenetics at this stage (Damn! I would have lived forever).

We have already covered the importance of curcumin and its role in epigenetic drift. Notably its ability to suppress genes that cause inflammation and to upregulate genes that suppress cancer.

Other great foods believed to influence epigenetics are broccoli, kale, cabbage, watercress (because of their isothiocyanates or sulforaphane content). Folate which is found in beans, grains and other green vegetables is also a great epigenetic modifier. Selenium rich foods such as nuts, legumes, beans and some meats also have been shown to influence the epigenome. Meat is a hard one to advise safely though because of the science surrounding meat and aging, but also because of the steroids and antibiotics which is given to livestock and poultry.

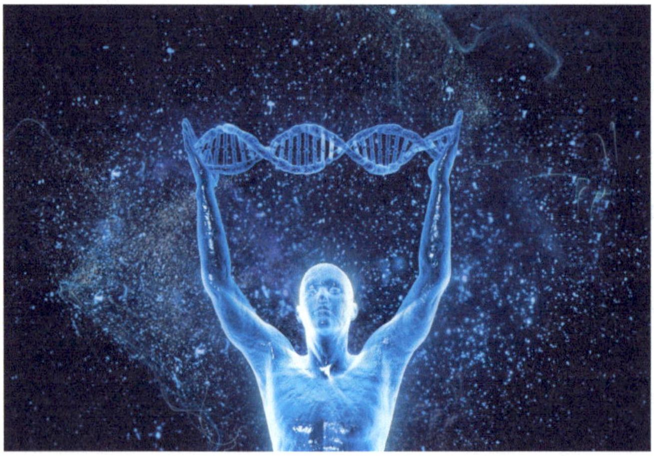

There has never been more evidence to say that we are what we eat. Our genes behave based on what we eat and the evidence also points to parents being able to preprogram the epigenome of their offspring, predisposing them to a life of obesity even though the genes remain the same. Your lifestyle may echo down through the generations.

It is nurture over nature, meaning your lifespan is controlled by how well you take care of yourself; your genes are not your destiny.

You hold in your hands how your genome behaves, so take care of it (my paper below).

www.pubmed.ncbi.nlm.nih.gov/34980713/

Just hit the View PDF for the full text.

The mTOR Pathway

The mTOR pathway is a biological pathway of various molecules that put you into growth mode. This is great for body builders, but it is also implicated in accelerated aging. The opposite of mTOR is raising AMPK from fasting and other techniques, which we have already looked at.

mTOR Inhibitors are a class of drugs designed to block the mTOR pathway which is responsible for various cellular metabolism, signaling and growth. This class of drugs shows great promise with suppressing tumor growth and aging (note we have already taken a glance at mTOR inhibitors earlier in this book under the section Metformin which also touched on Rapamycin). Rapalogs (Rapamycin and its analogues) are currently being explored as gerosuppressors.

- Gerontology is the science of aging, so gerosuppressors are drugs that suppress aging.

mTOR is a protein that signals other proteins that tell cells to grow and divide, namely cancer. Inhibiting that pathway with signaling inhibitors such as rapalogs (or even other mTOR inhibitors such as Everolimus) will prevent further signaling, which then prevents further cellular division, which in turn prolongs the time for you to reach the end of your cellular divisional limit (keeping you young).

mTOR functions to regulate cell growth and proliferation, cell motility and subsistence, protein synthesis, autophagy (which you will read about soon) and transcription (the reading of genes for mRNA>protein synthesis).

mTOR encourages the activation of insulin receptors and insulin like growth factor 1 (IGF-1) receptors, which again, puts you into growth mode by speeding up cellular division. We have already covered IGF-1 which is activated by red meat (and high protein diets). Remember, your cells can only divide a certain amount of time, once they hit that limit, they can no longer generate new proteins for your skin and body, or resist the onset of frailty (loss of cells is one of the biggest reasons you will enter frailty earlier or later in life, so hang onto them whilst new telomerase activators are under development). Alternatively, lift weights and exercise as that clearly boosts telomerase which extends telomeres and your cellular divisional limits.

You need mTOR activation early in life during development, this is the time it should be turned on, however as you age, you need to turn it down or even transiently turn it off. Leucine and methionine are two amino acids that are implicated in turning on mTOR, so sometimes reconsider foods which are rich in those amino acids, such as tuna, salmon, shrimp, beef, lamb, brazil nuts, soybeans, tofu, beans, lentils, wheat germ, chickpeas, brown rice and eggs on the days you want to either lower or turn off mTOR. You don't need to avoid those foods all day every day, just be mindful of what activates mTOR.

mTOR stands for Mammalian Target of Rapamycin.

So, let's take a quick look at Rapamycin.

Rapamycin – mTOR Inhibitor

Anti-Aging Score 9.5/10

Rapamycin has been implicated in lifespan extension and is also of course an mTOR inhibitor. Aged muscles that are experiencing sarcopenia (the loss of muscle cells) activate the mTOR pathway to try and build new muscle, however activation of mTOR does not lead to more muscle and blocks autophagy (tissue recycling), so even though rapamycin and metformin both block mTOR, which sounds like they will deliver us into muscle wasting and frailty faster, they allow for muscular regeneration by activating autophagy. Yes, biochemistry is a paradigm of strange logic.

Rapamycin was tested on the longevity of animals and even outperformed metformin; however rapamycin is an immunosuppressant possibly making the user more vulnerable to infections. True longevity requires a fully functional immune system, but rapamycin is increasingly being used by various longevity scientists and fanatics and many studies are underway. Rapamycin continues to open new doors of anti-aging enquiry and it is under consideration to be repurposed as one of the first geroprotection (anti-aging) drugs.

Rapamycin appears to be safe in animals, and of course in organ recipients who use it, but to a healthy person? It appears so, but I do not currently use rapamycin in my anti-aging toolkit until I can view more human data. That said, I currently use metformin which has a great safety profile and there are plenty of other products in this book I am sharing to keep you busy with. However, I must reiterate that rapamycin may be the most powerful oral supplement you can take, as the latest research is impressive, and I find myself leaning toward it.

For those of you that have an autoimmune disease, which are becoming epidemics in their own right, rapamycin is an immunosuppressant, so you would get the best of both worlds here. And rapamycin also protects aging muscle (the dosage for that was a short six-week program, at .15mg/kg of body weight). I don't have data on longer term effects, however the results are promising. There is also data that shows it may also encourage the body to burn fat over carbohydrates. I could really dive deep on the biochemistry at play here, but its too deep for a book like this, which I am trying to keep relatively light for you. Some of my readers love the deep science, but many of you just switch off, so it's a balancing act when writing a book like this.

Interestingly rapamycin was found in the soil on the Island of Rapa Nui (inside a bacteria known as Streptomyces Hygroscopicus). It is another naturally occurring compound that was already in nature just waiting to be found for what will most likely become part of your anti-aging toolkit.

Note that if you are already suppressing mTOR with Metformin then you do not need to also use rapamycin (both are prescription drugs and only your doctor can recommend the best one for you). Chances are that your doctor was not trained in the longevity technology discussed here, so print out some studies and take them with you when visit your doctor.

Rapamycin takes us onto our next topic, the FOXO genes.

Ref: Johnson, C, S, et al, Nature. 2013 Jan 17; 493(7432): 338–345
Ref: Zheng, Y & Jiang, Y Mol Cell Pharmacol. 2015;7(2):15-20
Ref: Aging (Albany NY). 2019 Oct 4;11(19):8048-8067. doi: 10.18632/aging.102355

The FOXO Pathway

FOXO genes help organisms protect and repair tissues. As we age, the ability of this gene becomes diminished and various tissues become damaged from wear and tear and lack of stem cell rejuvenation.

FOXO is a master control gene that controls many other genes that are responsible for keeping you healthy and young. Think of FOXO like the boss of a building company, below are the workers performing all the building functions such as concreting, installing doors, and ensuring the building functions and remains sturdy. Now imagine those workers are no longer being supervised properly, the quality will suffer, and in biology that usually means disease and premature aging.

Biochemist Professor Cynthia Kenyon has done some incredible work with the FOXO gene and is able to double the lives of 'c elegans' which are tiny round worms. What good are worms though? The genetic systems for longevity are incredibly similar in humans and worms.

Cynthia's research indicates that when times are tough or stressful such as rigorous exercise or lack of food (fasting), that the body signals the FOXO genes to kick in and ensure that all protective genes are turned on and functioning properly. Alternatively when things are good, such as an abundant supply of food, insulin growth factor 1 or IGF-1 (that we have already discussed and chastised earlier in this book) appears to alert the body through a range of mechanisms to turn down the FOXO genes.

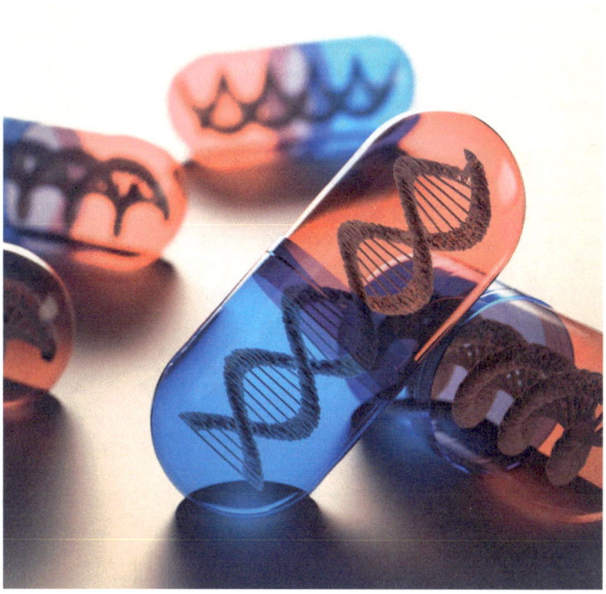

There are things you can do to stimulate the FOXO family such as fasting or exercise and of course use metformin; because metformin inhibits the signals that tell the FOXO gene that things are great (over eating) and assists with a greater calorie restriction effect.

Of course other great drugs are currently being designed to specifically upregulate (turn up) the FOXO genes, but for now the techniques shown in the paragraph above are your best bet.

Incidentally the name for the FOXO genes come from its actual tongue twisting name "forkhead box O genes', which is quite a terrible name to remember.

Aging Cell. 2016 Apr; 15(2): 196–207
PLoS One. 2014 May 23;9(5):e98207

The NRF2 Super Pathway

NRF2 (a protein which assists in transcribing other genes into proteins) is what I would consider a longevity regulator, meaning that it regulates detoxing the body and keeping us healthy. This is why you want to make sure you understand how to upregulate this gene.

NRF2 is instrumental in the phase II detoxification system, which is super important for onboarding fat soluble toxins and converting them into water soluble compounds for excretion. The phase II detoxification system also removes extremely reactive metabolites and other dangerous molecules from the body by neutralizing or inactivating them. Without NRF2 activation the phase II detoxification system would not function at full capacity.

So, let's see what else NRF2 does, and then I'll show you how to upregulate this biological pathway. NRF2 slows down or decreases various forms of inflammation and oxidation occurring systemically in the body. NRF2 is activated when the body reaches a certain level of oxidation. For example, the oxidative stress from exercise activates the NRF2 pathway, which of course is another reason why exercise is so important. This is also a form of hormesis, which you'll learn more about later.

Normal NRF2 activation can assist in aging, autoimmunity, arthritis, depression, kidney and liver dysfunction, neurodegeneration, drug exposure, exercise and microbiomic dysfunction, skin disorders and vision.

But NRF2 over-activation, which basically means turning it on too much, may aggravate symptoms for atherosclerosis, chronic inflammatory response syndrome, heart transplant, hepatitis C and Nephritis.

NRF2 is a protector, and when it senses various toxins, it can switch on to prevent damage to tissues. For example, cigarette smoke can turn on NRF2 in the lungs to protect the lungs from the dangerous chemicals found in that smoke. Please do not smoke to upregulate NRF2, that would of course be extremely misguided. And even allergens, viruses and bacteria entering the lungs can initiate NRF2 activation to protect lung tissues from damage. Even brain and liver can be protected from excessive alcohol consumption, but again, don't drink excessive amounts thinking that this is a healthy approach to upregulating NRF2, because it isn't.

So how do we upregulate this little transcription factor? Well, calycosin which is found in Astragalus can activate NRF2. Piceatannol, a resveratrol analogue can also upregulate NRF2. Sulforaphane which is converted from glucoraphanin, comes from broccoli and broccoli sprouts is a potent NRF2 activator. I use broccoli daily for this exact reason, and it is worth knowing that people that get sunburnt easily may have low NRF2 levels, and this was shown in a 2017 study that NRF2 is able to protect collagen in relation to ultraviolet radiation. So eat your broccoli sprouts especially if you live in a sunny country and want to maintain youthful looking skin.

NRF2 also protects us from Advanced Glycation End products or AGEs, which come from cooked meats and fatty dairy foods and other processed foods. And NRF2 is also able to protect the body against high levels of heat related stress. I'll finish up, and say that NRF2 increases mitochondrial function, which means more energy for your cells, and this effect can be increased with cordyceps, CoQ10, including another substance called Hispidulin. There are of course other ways to raise NRF2 such as calorie restriction, alpha lipoic acid, mucuna, Tribulus and even ashwagandha are all great candidates.

Senescent Cells and Senolytic Drugs

As we age millions of cells become dormant and linger in the tissues amongst our healthy cells. These dormant cells are known as senescent cells and can cause a myriad of trouble for us. Instead of dying through programmed cell death which is known as apoptosis, these senescent cells can begin sending out signals to the surrounding cells and tissue which cause problems such as inflammation and even disease.

When the number of senescent cells becomes ubiquitous amongst our tissues, it affects many other cells adversely and speeds up the effects of aging and propensity for disease.

Luckily for us all, scientists have developed weapons known as "senolytic drugs" to clear our bodies of these cells to ensure our tissues are rid of these trouble making cells.

There are many moving parts inside a cell and just like any system with moving parts, if those parts do not work flawlessly then incorrect function can creep in.

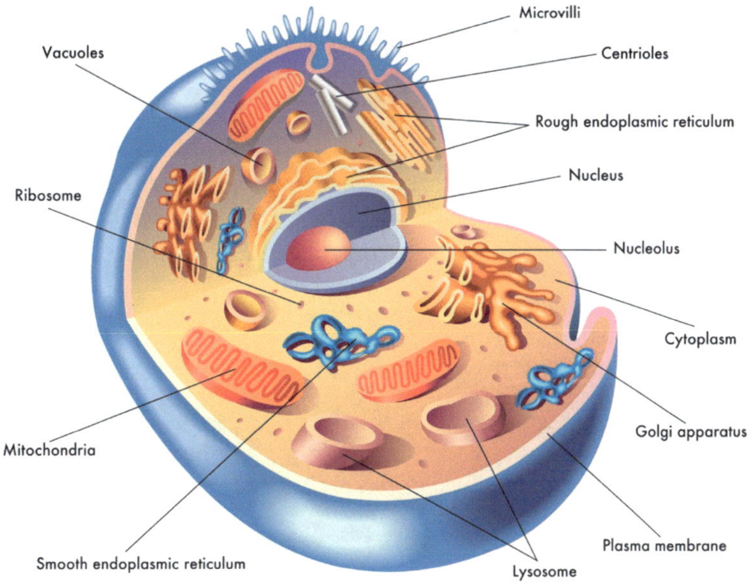

A well-known senolytic compound that has shown great promise is Fisetin which is found in fruits and vegetables. This compound was discovered in a recent study conducted by the University of Minnesota by Professor Edgar Arriaga.

Fisetin also inhibits an inflammatory response in human osteoarthritis by activating the SIRT1 gene (which is one of your primary longevity genes). A study published in Advances in Experimental Medicine and Technology in 2016 says -

"Fisetin (3,3',4',7-tetrahydroxyflavone) is a dietary flavonoid found in various fruits (strawberries, apples, mangoes, persimmons, kiwis, and grapes), vegetables (tomatoes, onions, and cucumbers), nuts, and wine that has shown strong anti-inflammatory, anti-oxidant, anti-tumorigenic, anti-invasive, anti-angiogenic, anti-diabetic, neuroprotective, and cardioprotective effects in cell culture and in animal models relevant to human diseases."

The fact that Fisetin is already a senolytic (something that can clear senescent cells), but also has so many other benefits such as those listed in the paragraph above, means that Fisetin would be hard to leave out of our anti-aging toolkit.

And note, you can get Fisetin from natural foods as shown in the quote on the previous page. The evidence to eat plants continues to come in as this book is researched and written. I am yet to read a study where it was better to consume dairy or meat products than plant based foods.

I simply go where the scientific evidences leads me, and so far, the evidence is overwhelmingly in favour of a plant based diet.

Another senolytic compound (but this time pharmaceutical and much more powerful) is known as dasatinib. This senolytic drug is used to clear cancer cells and should only be used with extreme caution and under the watchful eye of a professional medical practitioner that understands how the drug works.

I am a user of senolytics and will outline later in this book a regime on how to implement senolytics and many of the other tools you should have in your anti-aging toolkit by now.

- Apoptosis is programmed cell death.

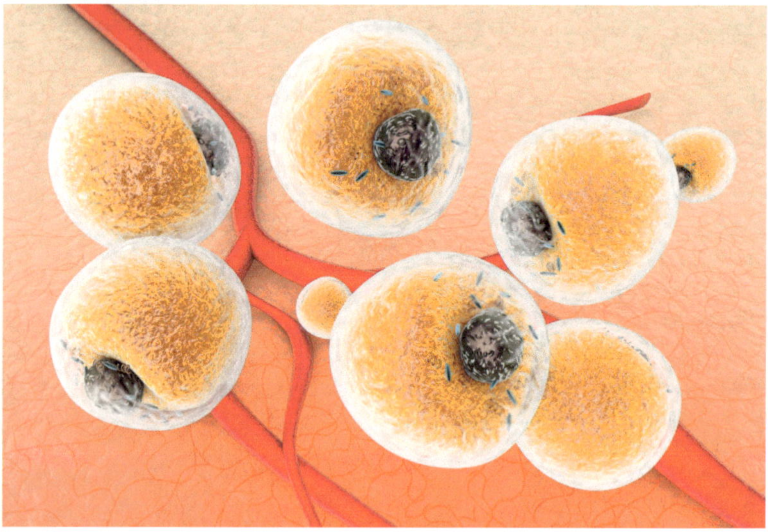

Some of the FOXO genes previously discussed have apoptotic pathways as well, which means they are responsible for programming a cell's death. This translates to FOXO clearing out senescent cells that do not die and just want to linger and send out inflammatory signals to the surrounding tissues. Senescent cells are a major cause of premature aging and disease.

Clearing out your bodily tissues of disease causing senescent cells is one of the most powerful anti-aging pathways you can harness as it lessens your body's chance of premature aging and disease.

Nature. 2014 May 22; 509(7501): 439–446
Nat Med. 2015 Dec; 21(12): 1424–1435
Nat Rev Drug Discov. 2017 Oct; 16(10): 718–735

Autophagy and Lipofuscin

A process that your body must undergo to stay young and steer off disease is autophagy. Autophagy is the recycling of cellular components that may have become redundant. Your cells can actually re-use and recycle proteins and other worn out cellular parts so that the body becomes more efficient at using whatever resources may be available to become more energy efficient. In fact, amyloid proteins that can accumulate inside the brain to causes Alzheimer's can even be recycled through autophagy and made into useful biological machinery.

If cellular junk is not being processed or removed from your cells, you end up with senescent cells misfiring and causing all types of localized problems in the surrounding tissue.

The million dollar question is, "how do we activate the autophagic pathway?"

Fasting is the first way to achieve this. And then suppressing the mTOR pathway with metformin or rapamycin is another good way. These two techniques combined are a powerful way to activate autophagy and ensure that cells do not become senescent in the first place.

Remember that metformin is a calorie restriction mimetic which helps trick your body into thinking it is fasting.

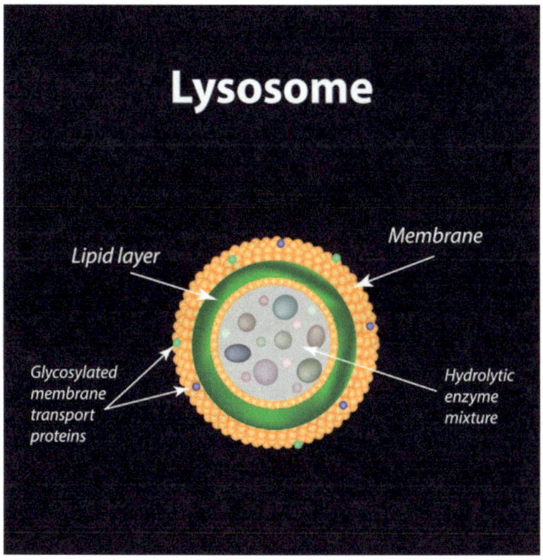

It is worth knowing that the lysosome is the cellular machinery inside the cell that is responsible for recycling waste proteins and products into useful parts.

Lipofuscin is the name given to this cellular junk that goes on to inflict damage on our cells, causing them to go senescent and ultimately damage the surrounding tissues. Lipofuscin may consist of poisons such as metals which may include aluminum, copper, iron, mercury, zinc, etc. Lipofuscin may also contain sugars and lipids among others. Many of these things can be recycled into something useful, but if they sit dormant inside the cell they can cause major problems as well.

The concept of recycling has been happening for billions of years and takes place millions of times a day inside your body. So next time someone asks you to recycle that plastic bottle by using the recycling bin, know that this is an important mechanism for life to exist in this universe and a good deed worth doing.

Mol Cells. 2018 Jan 31; 41(1): 45–49

Skin Aging – The Skin-o-Meter

Because your face and skin are the first thing people see when they look at you, it is evident that your face and skin quality are a human's first measurement gauge for age, otherwise known as the "skin-o-meter".

However there are many things you can do to remain smooth and wrinkle free. Probably the most obvious insult to skin is sun damage. Keeping UV free maintains a younger looking skin. Lifestyle factors are also big factors in skin health such as smoking, diet and exercise. Note that a poor diet and lack of exercise leads to an increase in free radicals and oxidative stress, which speeds up wrinkle formation in the epidermis (skin).

The skin is our largest organ and is designed to keep things out such as bacteria, viruses and other types of poison substances. It does a great job at this and also functions as a thermoregulatory device (sweating cools you off).

Many of the spots you see older people develop later in life (sometimes called liver spots or sun spots) are the result of free radicals, sun damage or over expressed melanocytes (pigment cells). These are caused by cellular dysfunction including senescent cells, lack of autophagy or a buildup of lipofuscin, and also lower levels of enzymes and anti-oxidants that decrease with age, etc.

Skin elasticity and density (loss of fatty tissue under the skin) are also affected as we age. In fact because of vascular aging, blood vessels become smaller and thinner which means blood flow carrying essential nutrients are restricted to the cells and bruising becomes easier. To add insult to injury, because of reduced blood flow the time it will take to heal becomes longer as well.

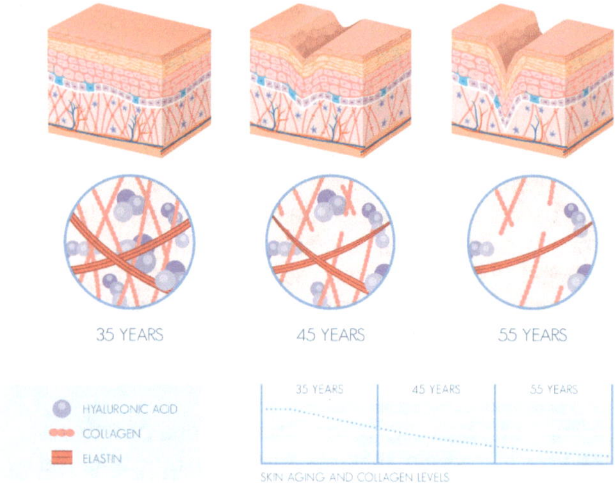

The skin is made up of many products but importantly collagen, elastin and hyaluronic acid all wane as we age.

The sun can also break down the elastin fibres inside the skin which will cause the skin to stretch, droop, sag and prevent it from staying taut much faster than the natural aging process.

However we already know how to slow or stop a large part of vascular aging with NAD and an epigenetic diet which assists in expressing the Sirtuin Genes with epigenetic foods such as curcumin, resveratrol and green tea among others.

The best time to start looking after your skin is as early as possible. Young people think they are immortal and will always be beautiful, the fact is, that every time a young person gets sun burnt, even though that damage may appear to heal and disappear, those damaged cells are in fact releasing all types of inflammatory responses that damage the surrounding tissue which can come back to haunt their pretty looks years later. So the earlier you take skin care seriously the better.

Supplementing with longevity products such as NAD, Arginine, branch chain amino acid complexes and telomerase are shown to attenuate the causes of skin aging. The epigenetic and plant based diets are also paramount to ensure rejuvenation of skin takes place efficiently.

I am an inventor of biomedical products and has a worldwide patent on a particular skin cream which delivers specific enzymes directly through the stratum corneum (top layer of skin) and into the skin cells that reside between the epidermis and dermis. This is mostly done with technology you have already read about in this book. The cream delivers telomerase (among other things) to the skin cells beneath so that the cells can divide indefinitely (preventing you from reaching your skin cell divisional limits and so that skin cells may function more youthfully). When it comes to skin cells, you can spend all the money you want on fancy creams, but if you are running out of skin cells or the cells are failing to repair the skin properly, then you are basically wasting your time (and money). Even though this seems like a shameless plug to promote my own technologies, it is actually my job here to keep you abreast of what technologies may or may not work. I could spend .000001% less time and effort that is required to write a book like this, and simply use social media and reach a million more people each night to promote my skin cream, so no this paragraph is not written to promote a skin cream. But hopefully this technology will be a game changer for those that want fresher looking skin. The cream is still under development and therefore we do not have before and after photos to show you but we will certainly publish those as soon as they become available.

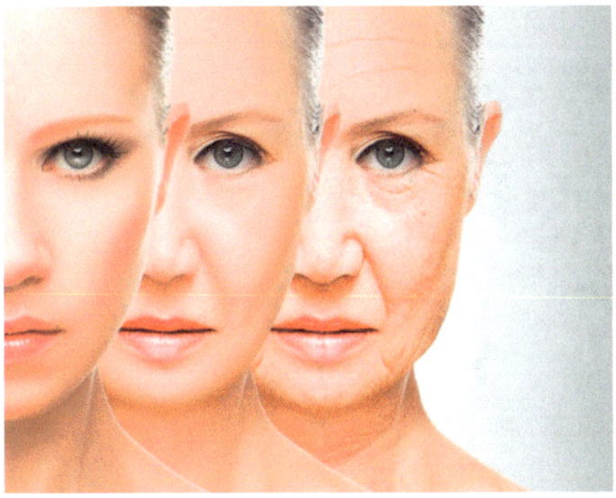

Older people's skin is also a weaker barrier against infections, and coupled with a weakened immune system that comes from age, this is also another prime reason why taking care of your skin's health as soon as possible is so important.

The Center for Genomic Regulation in Barcelona has drawn the conclusion that skin cells known as fibroblasts that manufacture proteins including collagen lose their identities as we age. This results in cells sitting inside the epidermis that no longer know what to do to keep your skin looking full and tight. Even though there are numerous other reasons the skin can look weathered, it is the genetic expression (the epigenetic drift) that are the primary causes of this condition.

So is there a way to prevent your genes from drifting into the old age state? Yes, of course there is, with the epigenetic diet. However even better Dr Brad Thompson from Calgary, Canada is designing a skin rejuvenation program with gene editing; he wants to be able to reprogram your DNA by upregulating genes which control the appearance of skin. I say to people very regularly "get ready to live inside a science fiction movie, these technologies are almost here".

Cold Spring Harb Perspect Biol. 2013 Jun; 5(6)
Exp Dermatol. 2011 Apr;20(4):297-302

The Klotho Gene / Protein

According to the Human Molecular Genetics Journal there are about 19,000 to 20,000 genes inside the human genome (November 2014). We are only interested in genes that may affect aging here though. A gene that is currently receiving a lot of attention is the Klotho Gene.

The Klotho gene is able to upregulate two other genes known as FGF21 and FGF23 which help rejuvenate your body. Klotho does much more such as heightens brain function, removes harm caused by oxidative stress, protects the kidneys, protects the cardiovascular system, rejuvenates muscle, subdues growth pathways such as mTOR and is shown to extend life span in various organisms. It is worth knowing that Chronic Kidney Disease has now been strongly linked to low levels of Klotho.

As with all things that keep us young, the levels of Klotho inside the human body drops off as we age which means this is just another pathway of aging which we may be able to control.

Klotho is now well regarded as an anti-aging protein that seems (at first) to defy logic. Insulin which is released by the pancreas into the body when we consume sugars seems to elevate the Klotho levels. However before you rush out and purchase a carton of soft drink, know that Boston University's School of Medicine performed a study on this and found that Klotho was actually released to suppress the effects of the insulin on the cells.

So one of the ways that Klotho appears to suppress aging is by attenuating the effects of insulin and insulin growth factor 1 (IGF1, which we have discussed earlier in this book). If you are eating a vegetarian based diet, and steering clear of sugars and suppressing growth pathways with drugs like Metformin, then you are most likely already doing what Klotho does with insulin.

If the Klotho protein can be synthesized so that you can enjoy it without high insulin levels then you are in a prime position to slow down or block many aging pathways.

Klotho is named after the Greek Goddess who is believed to have spun the web of life.

A recent study published in the Journal of Gerontology has found that a Klotho deficiency also restricts the production of telomerase which in turn shortens telomeres and life expectancy. The study demonstrated that when Klotho levels were low there was impaired differentiation of stem cells into other cell types, cellular senescence and stem cell death (apoptosis).
Ullah & Sun 2018, The Journals of Gerontology: Nov, 2018

Increase Klotho levels with sulforaphane (broccoli), green tea, cordyceps, omega 3's and a GABA supplement (Gamma-Aminobutyric Acid) or scan the code.

Thymus Regeneration

The thymus gland generates T-cells for immune function. T-cells are a type of white blood cell and are imperative for a healthy immune system. As we age the thymus gland shrinks, generates smaller amounts of T-cells to fight infection until your thymus gland becomes so weak that you can no longer fight off serious infections or diseases such as cancer. This is why we see older people die from common conditions such as the common flu (it is also one of the reasons why older people get more disease such as cancer). This condition of a weakened immune system is often referred to as immune senescence.

The thymus in a child is quite large compared to a full grown adult, and things get worse as you enter old age. The thymus gland sits between your lungs as shown in the image below (child on the right).

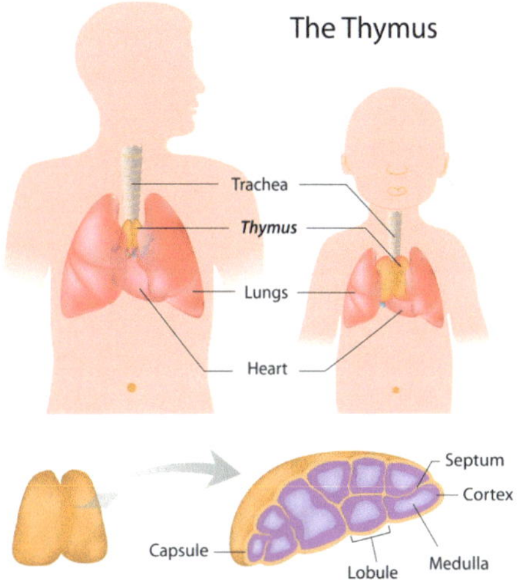

If you are leaving middle age it is time to take your thymus seriously. Technologies are currently being developed to regenerate the thymus gland. Scientists have recently found that if you take thymus progenitor cells (similar to stem cells) or a growth factor protein known as BMP4 and inject one of these into the bloodstream you can regrow the thymus (in animal models at the moment). This also translates that a new thymus gland could be grown in the lab and transplanted later. This method is invasive and no doubt will be costly (for a government it could actually save billions of dollars by allowing older people to fend off disease that would otherwise hospitalize them).

As a reader of this book you may want to know what therapies are available that do not involve regrowing or transplanting a brand new thymus. Can we reinvigorate the existing thymus? And yes, it looks like we can with some easily obtainable products.

Before we look at those compounds though, it is worth knowing that a lack of immune function is directly correlated to an increase in senescent cells. Basically without immune system surveillance dysfunctional cells can run rampant causing a cascade of tissue dysfunction and disease.

Immunity is also increased when we suppress the mTOR pathway with products such as metformin or rapamycin. Note that downstream of the mTOR pathway exists the TORC1 protein complex which is shown to have an inverse effect on immune function (meaning suppress TORC1, increase immune function). So the compounds I am recommending in this book such as metformin are rarely for a singular path, but suppress multiple aging pathways.

- Metformin > Suppresses mTOR > Suppresses TORC1 > Upregulates Immune Function > Stronger Health > Less Risk of Disease > Longer Life > Yeehar!

Dr Greg Fahy is leading the charge in this field and is exploring novel therapies worthy discussing. Using human growth hormone (HGH) and dehydroepiandrosterone (DHEA) Dr Fahy has seen a distinct enlargement of the thymus. And even more recently another study with HGH, DHEA and metformin was found to reverse the methylation markers that are found in the epigenetic clock. Some would say that this equates to age reversal.

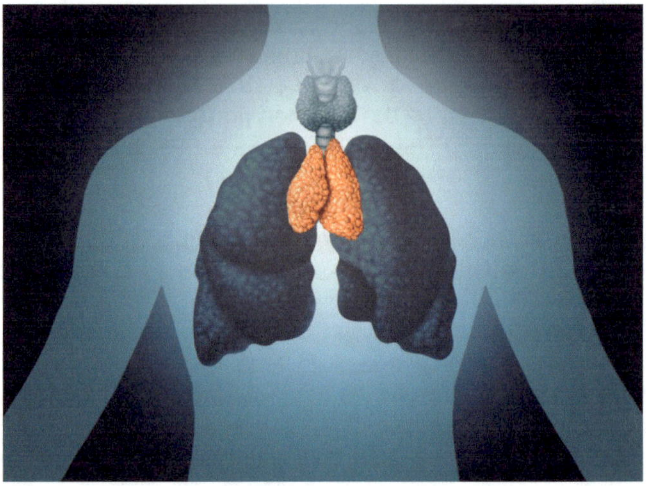

You can stimulate the immune system with the above, and easily with three products (below) that do not require a prescription (and there are other products of course, but these combinations above and below stand out from the crowd).

1) Reishi mushrooms
2) Cistanche
3) Pu-erh tea

Reishi mushrooms are shown to have anti-tumor effects by increasing immune function. Interleukin 6 (IL6) is a pro-inflammatory compound and has been connected with early mortality; however these mushrooms are shown to lower the levels of IL6.

Cistanche is a plant found in the desert and was also shown to reduce levels of IL6 whilst boosting immune cells in the body. Animal models demonstrated a 15% longer life span with Cistanche.

Pu-erh tea is shown to increase the amount of T-cells which prevents immune senescence. This clever tea also slows down bacterial and viral growth. And yet again was shown to reduce IL6 by 21%.

All three of those are standouts and if you have a weakened immune system from therapies such as chemo or radiation or simply have a hard time recovering from infection then you should consider this information for your anti-aging toolkit.

Babu PD, Subhasree R. The sacred mushroom Reishi-a review. American-Eurasian Journal of Botany. 2008;1(3):107-10.
Fahy J Anti Aging Med. 2003;6(3):219-27
Zhang YH, Wang LC, Tu PF, et al. Macrophage activation by low molecular weight saccharides from Cistanche deserticola. Zhongguo Zhong Yao Za Zhi.2017 Nov;42(21):4207-10.
Lee LK, Foo KY. Recent advances on the beneficial use and health implications of Pu-Erh tea. Food Research International. 2013 10//;53(2):619-28.
Ovadya Y et al Nat Commun. 2018 Dec 21;9(1):5435.

Hormetic Theory aka Hormesis

Interestingly that which would kill you may be healthy, and that which is healthy may kill you.

Hormesis is the body's response to certain stimuli. Some of that stimulus in small doses could be healthy, but in larger doses could be harmful. Let's take a look at how our bodies work.

Logic may suggest that lying on the couch day after day and eating as much food as possible would make us healthier. After all we are not using our body and wearing it out, we are eating as much as possible and maintaining as much fuel and nutrients as our body can handle. So at face value it would not be far-fetched to think that this type of lifestyle would extend life. However we know the opposite is true. Exercising and fasting extend life, but why?

When we exercise our bodies undergo extreme amounts of oxidation (the release of free radicals) which can actually be harmful to our bodies. Luckily for us, our body also has a response to this in which it releases vast amounts of natural anti-oxidants (catalase, super oxide dismutase, glutathione, etc) which purges not just the free radicals from the exercise but also the free radicals that may be lingering in our bodies simply from lifestyle. Other beneficial facets such as stem cells are also released to repair the joints and muscles, and our telomere attrition is also attenuated (slowed down). This means that it is not the exercise that is actually healthy, but our body's response to that exercise that is healthy.

Another example is that green tea extract in large amounts can be harmful; however the polyphenols inside green tea when taken in small amounts can be extremely healthy. Again it is the body's response in purging the harmful effects of polyphenols that is healthy, and not entirely the polyphenols (polyphenols are also antioxidants and perform other functions, but some aspects can be harmful). It is safe to drink three or four cups of green tea per day, so do not fret.

This type of negative stimuli which turns into positive stimuli is known as a biphasic response in biotech circles.

Worth mentioning that Paracelsus, a Swiss born alchemist from the 1400's was the first to note that compounds in large doses that were harmful, could be helpful in small doses. His direct quote is –

Sola dosis facit venenum (Only the dose makes the poison).

Paracelsus is correct hundreds of years later, he was a futurist way ahead of his time (as all scientists / alchemists, early astronomers, explorers were back then).

Hormesis is a weapon you should use to extend your longevity, just be mindful of the dose and effects.

Examples are sunlight, standing instead of sitting at work, ice bath, stairs instead of escalators, etc. Small doses of stress work.

Conventional Medicine - Vaccines (Type IV)

There are many vaccines for things like hepatitis, meningococcal, human papilloma virus (HPV) and influenza etc. Now you would not argue with a rocket scientist at NASA about how best to launch a rocket, unless you had done rocket science. And you would not argue with an astrophysicist about how stars form unless you were highly literate in astrophysics. So, please do not argue with an immunologist (unless you are one) about vaccines being dangerous because you saw something on Facebook or because Bob the plumber next door said they were bad. Vaccines are a biological shield that will protect you from a myriad of preventable diseases, so start vaccinating if you are serious about longevity. Why would you leave those viral doorways open as you enter old age?

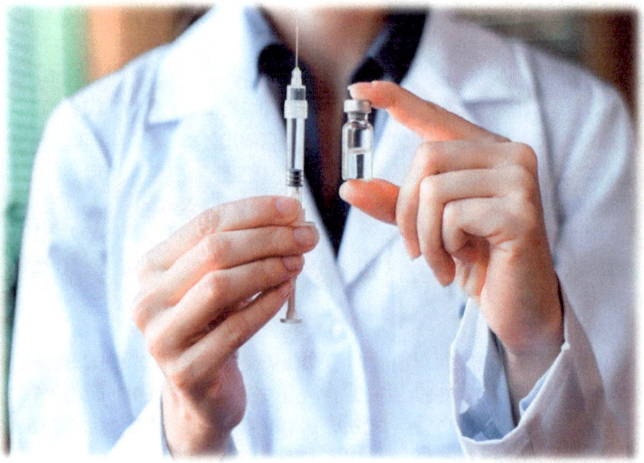

Some of the misinformation around vaccines is that they contain formaldehyde, mercury or aluminum. Let's take a quick look at those claims of toxicity.

Formaldehyde - of course drinking formaldehyde is going to be bad for you, but remember what we just said on the previous page, 'only the dose makes the poison'. Formaldehyde is used to disable proteins and to eradicate viruses during vaccine preparation. There may be microscopic trace amounts of formaldehyde in vaccines and those trace amounts never amount to more than .1 milligram. An apple contains six milligrams (sixty times more) and other plants and fruits contain just as much. You have probably even eaten cyanide (yes cyanide) that also comes in some fruit, so do not let your lack of scientific knowledge be exploited by fear mongers. It is the dose that makes the poison.

Mercury – Most vaccines do not contain any mercury. The type of chemical we are talking about here is known as thiomersal (ethyl mercury) and is only really used during an epidemic where biotech companies cannot make single use vials quick enough (thiomersal is used to sterilize and decontaminate vials for its antibacterial properties). The amount of mercury you would get from a serve of tuna is around 85 micrograms and would stay in your body for much longer (around fifty days) because mercury from fish is methyl mercury, whereas ethyl mercury (thiomersal) lasts only seven days in the body. Of course mercury is found throughout nature and in your diet at much higher doses already, so this discussion truly is 'much ado about nothing'.

Aluminum – Most vaccines use less than .1 milligram of aluminum as an adjuvant (which improves the immune response). Human breast milk contains up to 40 micrograms of aluminum per litre and baby formula even contains around 225 micrograms per litre. Aluminum is removed from the body after 24 hours (it is easily excreted by the body). You can safely consume up to 100 milligrams per day through diet, so .1 milligram (1000 times less) is well below any dangerous or toxic threshold.

Doctors / Scans / Tests (Type IV)

All general health is a form of longevity, even the act of washing your hands is a form of life extension. But we can go much deeper with regular check-ups with our doctor and regular scans (MRI and Ultrasound) to look for early stage tumors or worse. You'll need to ask for those types of scans because as already mentioned earlier in this book, current medical logic is reverse engineering, you develop a tumor with complete surprise, which now poses serious health problems including impending death, at which point your doctor tells you to get a scan. Catch those things early and you'll live much longer. If the doctor thinks pre-emptive scans are a waste of time, get another doctor.

Also blood and saliva tests can detect a myriad of biomarkers which can also assist in preventing a plethora of serious disease, so testing two or three times a year would be very efficient in ensuring you catch those things early on.

Second Opinions

Always get second opinions when dealing with life threatening diseases or diagnosis. Try not to let the second Doctor know you already have an opinion. This is not because your first Doctor is bad, but because we are dealing with your life. Third opinions are even better.

DNA Testing (Type IV)

DNA testing is now "run of the mill". It is cheap, easy and everywhere. Be proactive and go after the knowledge that tells you what disease you have a predisposition to. The shock of hearing this information will probably be enough for you to start taking steps to make sure your biology is setup to prevent the onset of whatever that disease may be. Much more on that later though.

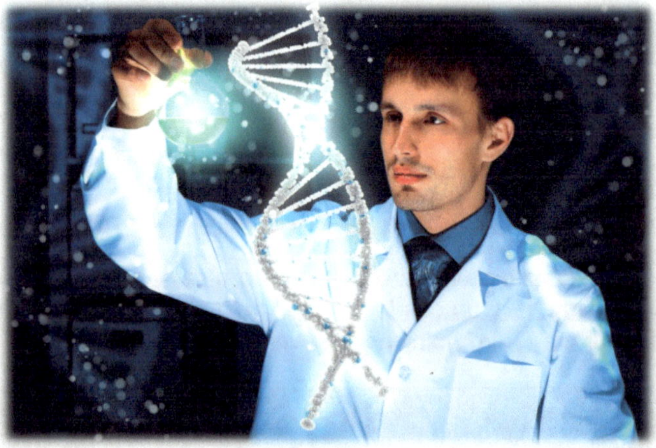

Reality Check - Whatever your predisposition, a plant based diet is front and center every time for the lifestyle change that will ensue, so some of you may have to toughen up. Let's be real, you are not ever going to hear your doctor say "eat more sausages" and you are not going to read a scientific study that says "the evidence shows that you should gorge yourself on more potato chips if you want to ward off cancer". So do you even need the DNA test if you already know the lifestyle change to make?

Genetic Disorders

Some other exciting science that is taking place is the investigation into genetic disorders such as Loran Syndrome and Amish Genetic Disorder. These two disorders give those that have it incredible protection against disease such as cancer, heart disease, diabetes, and that's even if they are obese. These genetic disorders seem to make those that have it immune from all the disease that everyday humans like us are destined to die from. But how? Scientists are working very hard right now to pinpoint the exact genes and mechanisms so that we may be able to supplement and hold this unique immunity from the world's most devastating diseases.

Nootropics

Smart drugs or cognitive enhancers are known as nootropic drugs. There is a growing industry in this sector and Silicon Valley is one of the biggest markets for nootropics. Some nootropic drugs are legal concoctions of various enzymes and chemicals whereas others are highly illegal. Known as micro dosing some of these harder nootropic drugs are taken in tiny doses so not to alter a person's state of mind, but enough to allow super complex mental processes to take place during the work day. This is certainly not recent either, it has been happening for decades, without incident but with powerful results (Nobel Prize winning even). Nootropics will cross over with anti-aging drugs in the coming years.

The Yamanaka Factors - The Four Genes

Shinya Yamanaka is a Nobel Prize winning stem cell scientist and researcher from Kyoto University in Japan. He won this prize alongside Sir John B. Gurdon for discovering that mature cells that had already become a specific type of cell could be reprogrammed back into a pluripotent state (a stem cell ready to become any other cell type except those cell types that form inside the amniotic sac and placenta). It was previously thought that cells could only stay in their differentiated state once they had reached it, or that they were unidirectional, how wrong the naysayers were.

So, what do you 'the reader' think that this technology holds for anti-aging?

Known as The Yamanaka Factors these four genes are known as the Oct3/4, Sox2, Klf4 and the c-Myc genes. By being able to reverse cells back into a pluripotent state and enable them to become any other cell type, this means that any cell types can now be made to repair other parts of the body. They can be used to repair the spinal cord and also used to repair brain injuries such as stroke, muscle and bone after accidents, even repair a damaged heart muscle after a heart attack, the list goes on. These cells would be taken from the patient that requires them, meaning there are no ethical issues with embryo's and because they come from the same patient the chances of rejection is extremely small.

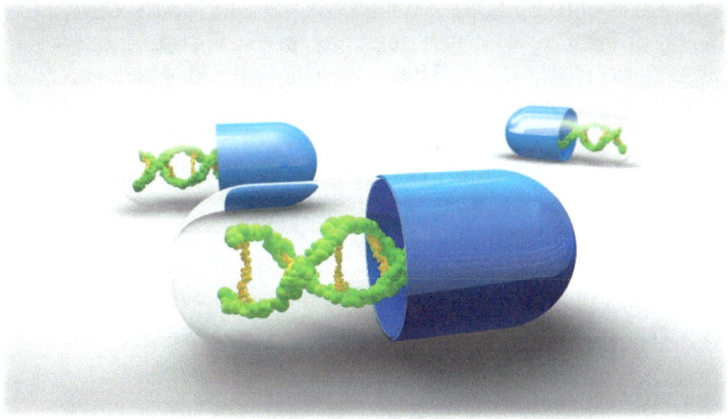

So what does this mean for you? Well performing this technology is relatively simple for someone literate in the science. Want fresher looking skin? Want to regrow your auditory nerve from age related hearing loss? Want to repair your eyesight? The list is endless and the stem cells are already yours.

When the Yamanaka Factors (genes) are expressed (turned on) it instructs human cells to function as they did when they were youthful and vibrant. Imagine looking at your wrinkled hand and taking an enzyme as a supplement that turned on the gene expression in these Yamanaka Factors. It is feasible that within weeks of taking that enzyme your skin would return to a youthful state. Science fiction? Not so say the scientific studies that are already coming in from the Salk Institute for Biological Studies in San Diego.

All manner of disease and conditions are being studied (and some cured) by controlling the Yamanaka Factors and the results are nothing short of futuristic.

We've already discussed DNA Methylation in this book. The science of epigenetics is that during life some genes either become more active or less active which controls how cells function (cell activity regulators). By controlling the Yamanaka Factors, we may very soon be able to control everything related to aging (and in the near future as well). Of course there are challenges, nothing is ever simple, but the science and frontier for eternal youth is getting closer.

Mesenchymal Stem Cells (MSCs)

Mesenchymal stem cells (MSCs) are found pretty much everywhere in your body and control inflammation, regulate the immune system and stimulate regeneration.

Our stem cell numbers decline with age and this is why we do not see most elite athletes older than their thirties. Your MSCs begin to taper off around the age of twenty-four and from there physical ability also tapers off.

When you have a major health problem such as a heart attack or stroke your MSC pool takes a big hit as your body uses vast reserves of MSCs to repair the damage. If you have more than one major health episode, things become even worse for your MSC pool. So this is yet another reason to maintain a healthy lifestyle and prevent any major health event from happening in the first place.

Another facet of MSCs worth noting is that the ability of MSCs to repair parts of the body diminishes as we age through what is probably slower cellular division. At a fetal age a single MSC can divide into two, then four, eight and so on, and within a month you will have a billion MSCs ready for growth and repair (from that single MSC). The time taken for a MSC to divide (on average) is around eighteen or nineteen hours. However, when you are a young adult, it would take around thirty days to create just 32,000 MSCs (as opposed to a billion at fetal age), which does not fare well considering you are much larger than a fetus and probably require many more. This is exactly why kids recover from injury faster than adults, and adults faster than older people.

The point of all this is that MSCs are extremely important for youth and longevity. Dr Neil Riordan from the Riordan Clinic in Panama says that every chronic disease is either secondary to either not having enough MSCs or having MSCs that no longer work properly (not secreting the right stuff).

MSCs are also present in cord blood. By administering cord blood into humans who had almost reached their mesenchymal stem cell limit, you can reverse the mesenchymal exhaustion and put those people back into a rejuvenated state.

Note - Corticosteroids that are usually injected into sore joints as anti-inflammatory agents are toxic to MSCs. So if you have joint pain, get an injection of MSCs if you can, because that is the only thing that will repair you. Corticosteroids will take away your pain temporarily but also ensure your chances of repair are less.

Note - Exercise increases your capillary density which also increases the areas where MSCs can home in on. The better your capillary density the more MSCs you will have floating around your system, the better the blood supply network, so the faster you can repair and rejuvenate.

With telomere manipulation we can now have an infinite supply of MSCs and easily rejuvenate and repair ourselves wherever and whenever. It is just going to take a few more years for this to become affordable and mainstream, but for those hungry enough, it is available now.

Mind Uploading? Seriously? Yes, Seriously!

Mind Uploading believe it or not is fast becoming a very real project where someone near death can upload the entire contents of their brain onto a digital hard drive. Sure, the technology is many years away, but going digital may be a way to outlive and outperform a biological body.

The Connectome Project is right now trying to map the entire contents of the human brain as a lead up to this type of technology. Brain to brain communication is already possible and the uploading of neuronal information from mice to microchips is already being done. A digital robotic human would not require food or get disease or even need sex. Sure, where is the fun in that you ask? But you only enjoy sex because you are programmed by nature to enjoy it, take away the need to reproduce with robotics and sex becomes laughable for species that need to perform it for survival.

With mind uploading your conscious being would reside inside some supercomputer and you would be sentient and possibly control yourself through robotics.

Your consciousness is nothing more than electrons bouncing around inside a biological circuit called a brain, so why not bounce those electrons around inside a computer where you are not subject to disease and neurological decay?

But what if someone was uploaded to a supercomputer that sat on a desk somewhere? Would that sentient computer now have the same rights as you or me? Could it be switched off and on at will? Could the information be edited as to create a completely different version of your consciousness? In fact Artificial Intelligence (AI) is already destined to have rights and AI is not even human, so chances are that any human that is uploaded to a computer would have rights.

Without the chemical signatures we experience at every moment of every day how would someone be able to feel emotions if they had converted into a digital format? Chances are someone will invent something for that as well.

Though an artificial body could sense much more than us, it could see in gamma, x-ray, ultraviolet and infrared wavelengths, hear radio waves, feel magnetism, so the sensory experience could be much more fulfilling.

You may think this is complete science fiction, but truth is stranger than fiction. Those faced with certain death that have the financial means to use this technology will almost certainly upload their conscious being onto a platform for immortality. And if you can fit and connect that platform into an extremely life like artificial body that looks like you, then you are back in business and did not really go anywhere.

Without disease, the requirement for food and the hassle of life support systems for space travel, there is every chance that in the future biological beings will be seen as outdated and primitive and machines will rule.

The NASA Twins Study– Telomeres and Gene Expression

Mark and Scott Kelly are both astronauts for NASA and also twins. Known as The Twins Study one of the brothers Mark stayed on Earth, while Scott went to space, NASA compared their biology during this time.

NASA watched very carefully as Scott was rocketed up into space and resided on the space station for a year. As a snapshot of this data, NASA learned that astronauts on long missions will experience oxygen deprivation, nutrient shifts and increased inflammation which all changed Scott's gene expression.

- Gene expression is simply when genes instruct the production of various proteins to perform targeted functions. This can be a positive or negative process depending on the gene that is activated.

But the intriguing part was not that Scott's genes began expressing differently (even though there is some serious science to that) but that his telomeres became much longer in space, which would allow his cells to divide many more times than his brother's cells, enabling him to live longer into the future than Mark. Scott and Mark are both 54 years old and showed profound differences in physiology during this experiment, and with space travel being expensive (and dangerous) I am not about to include space travel into your anti-aging toolkit.

However, your anti-aging toolkit is more important than ever, because the evidence now shows that just being on Earth is aging us faster, so get that toolkit going and show some control over aging.

Transhumanism

What is transhumanism? Well, ask ten people and get ten slightly different answers. However I will average it out for you. Transhumanism is integrating biological form with sophisticated technology so that humans can live for extremely long periods of time and also do things outside of what would have only been achievable with biology. For example, a transhumanist may be able to unlock a coded door with his or her thumb. Why do I use that example? Because I interviewed a gentleman called Zoltan Itsvan for a television show in San Francisco. Zoltan is the former head of the American Transhumanist party (a very serious political movement). Zoltan was also a presidential candidate in the 2016 elections. Zoltan is a transhumanist and has a chip inside his thumb which can open coded doors.

Of course transhumanism goes much further than just a chip to open doors though. At the moment, we are already seeing the emergence of transhumanism with hearing aids, artificial limbs, pacemakers, dentures, contact lenses among many other lifesaving gadgets. However transhumanists (from what I have learnt) are not just interested in fixing parts of the human body that are no longer working properly, transhumanists want to see shops selling synthetic hearts or eyes, etc, that you can purchase and have replaced whilst your biological heart is still in perfect health. And even though you may find that creepy, it actually makes perfect sense. Why wait for a heart attack or until you are close to death before servicing your biological machinery? You replace parts for your car well before they become dangerous, such as your break pads or oil, so why not your body? If this technology is available, then transhumanists want you to use it. And no, it will not just be for the rich, the transhumanist political party makes things very clear when it comes to healthcare for all.

That's me on the left, Zoltan on the right.

Transhumanists believe that we will enter an era where we can integrate ourselves so much with technology, that we would have to call categorize ourselves as Posthuman Beings.

Note that it is not just your physical form transhumanists are targeting, but also your mind. They believe that technologies are also coming that can supercharge the intellect. These are not the ramblings of science fiction writers either, there are many projects already underway to do this (as we saw a few pages back (Mind Uploading).

Biological Equilibrium

Today we're going to tackle something a little different and use a term that is not really used in anti-aging or longevity, but I personally have been using it for a long time because I think it conveys a strong message. I am going to chat with you about something I call Biological Equilibrium.

So what is biological equilibrium? Well, it is a myth, biological equilibrium doesn't actually exist, but let's chat about this unicorn anyway and see how it works. Once you hit forty years of age, your body will never sit at biological equilibrium ever again. Meaning, your body is now either in a slow state of biological decay, or through diet, exercise and clever longevity technologies, your body becomes stronger as you age. But your body will no longer sit and remain strong and healthy without your help. And this is why biological equilibrium after forty doesn't actually exist, you either grow weaker and become more susceptible to health conditions, or you become stronger through various interventions making you highly resistant to the conditions of old age.

So get young or die trying!

It is common knowledge that as you age, that things like your testosterone and human growth hormone levels will taper off. It is also common knowledge that your telomeres will get shorter and epigenetic drift will cause certain genes to change expression, which may speed up the aging process. Many other regenerative elements that may be able to rejuvenate you, also taper off and this is one of the research areas that my biotech company is currently focused on. We will be doing the worlds first human trial on this technology, and I'll elaborate much more on this through my social media accounts in due course. Other serum elements in your blood that speed up pathological aging also begin to rise, and this is not good, so my biotech company researches technologies on how to maintain a similar biochemical condition to that of a healthy adult between 40 and 50 years of age. This is what my biotech company is currently doing, with our human trial taking a cohort of seventy to eighty year old's and restoring youthful tissue function back into their tissues, so that their tissues function like a forty or fifty year old again. Sounds crazy, but this is happening, and sooner than you think.

So because your body cannot maintain biological equilibrium by itself after forty and is designed to grow weaker, my advice to you, which is never medical advice, is to perform exercises that make you stronger. Walking the dog around the park won't work to resist biological aging. Yes, any exercise is better than no exercise, but if I was forced to choose a type of exercise to ward off aging, I would have to choose resistance training. That is, lifting weights. I lift four times per week, and I aim to lift the heaviest weights I can, and very important this next part, this is with at least ten minutes of warming up the muscle group I want to train.

Maintaining strength comes with many benefits. Firstly, you will absolutely feel younger when your body is stronger. You will naturally raise testosterone and human growth hormone to more youthful levels, among many other rejuvenating factors. The bigger your muscles, the more glucose will be vacuumed up from the blood because muscles have huge energy requirements, meaning you will lose weight faster. Of course, a good diet is extremely important, and I will say again that losing weight is 95% done in the kitchen. You cannot train your way out of a bad diet, so learn time restricted feeding and fasting. I don't eat before lunchtime or after 6pm, that is a 6-hour window I have to eat and 18 hours where I do not eat. You may find that hard at first so ease yourself into it should you choose to do it.

The Causes of Biological Aging

So why do we age? How many facets are there to biological aging?

There are many pathways to aging and some are still debated as to whether they are part of the aging process or not, but what we already know is the following –

The Pathways for Biological Descent

1. Mitochondrial dysfunction
2. Telomere shortening
3. Immune senescence (becoming more susceptible to disease)
4. Diminishing enzyme and antioxidant levels such as NAD, glutathione, etc.
5. Epigenetic drift - youth genes versus aging genes (loss of Klotho, Yamanaka, Sirtuins, etc)
6. Accumulation of advanced glycation end products (AGEs)
7. Rampant mTOR
8. Inflammation (nf-kb cascade)
9. The FOX01 pathway
10. Cellular & stem cell divisional limits (unable to repair or rejuvenate)
11. Vascular aging (restricted blood / nutrient flow to tissues)
12. Sarcopenia (loss of muscle, partly from vascular aging)
13. Increased generation of free radicals and oxidative stress
14. Accumulation of cellular damage and junk (lipofuscin)
15. Alkalosis (improper PH levels)
16. The AMPK pathway (improper cell communication)
17. Senescent cells / cell dysfunction inducing disease (lack of purge)
18. And of course lifestyle, how many biological insults you take during your life (smoking, injury, diet, infection, exercise, etc)

If you have been paying attention to this book, most, if not all of the aging pathways above can be strongly influenced or completely controlled. I also want you to view those pathways as the primary causes of disease, because all diseases related to aging can be traced back to one or more of those pathways above. I have intentionally left out diseases such as heart disease or neurological disease as pathways to aging as they are mostly a side effect of the aging process (listed above) and can be kept at bay with the science discussed in this book.

There are other pathways that are still under early investigation such as the Wnt, TERT, TFAM, etc, but these are so fringe and the science on their importance to aging or how to control them is still sketchy at best, so I am leaving many out. The purpose of this book is to get you in and out as quickly as possible with practical knowledge that you can actually use, and not bog you down with years of theory and scientific jargon. As new technologies emerge then the book will be updated and I will use social media to keep you informed.

Quick Guide – Step 1 - Longevity Planning #1

One of the best things you can do in your fight for longevity is to monitor where you are biologically. This means that before you start taking a myriad of longevity pills and implementing lifestyle changes, it would be handy to have as many biomarkers taken as possible. Everything from cholesterol, blood pressure, resting heart rate, vitamin levels, lipid profile, skin tightness, vision, telomere length, epigenetic drift, genetic testing to discover strengths and weaknesses, etc. The more biomarkers you can get about yourself, the more informed you can be on what longevity supplements to take in the first place and the more data you can ascertain later when you have your biomarkers checked again a few months later.

Low levels of vitamin D can cause all types of problems such as weak bones, however if your levels of vitamin D are too high then this can also cause many problems (including weak bones) such as lowering your vitamin K2 levels. So if you can establish a longevity baseline prior to starting your regime, this would be very helpful.

I conduct about three biomarker tests per year. I would even recommend weighing yourself in the morning on a regular basis and monitoring how your weight may fluctuate daily with different foods. Note even taking a facial photograph every month may indicate when you are starting to rejuvenate (yes it's that easy to notice longevity techniques paying off). In fact it is worth stating that the number one thing people use to gauge another person's age is their face.

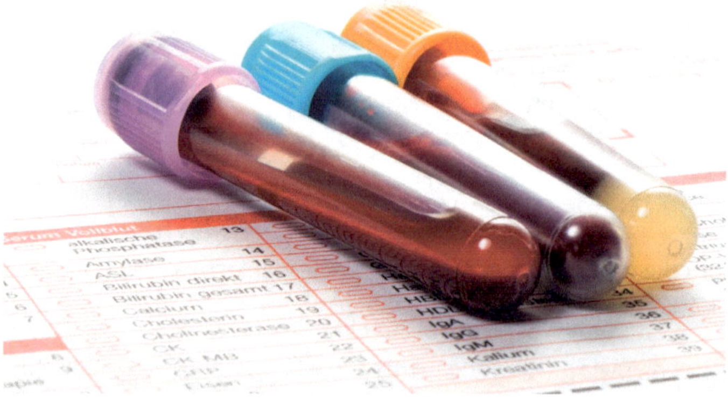

I must stress that everyone is different, that is that everyone's genes are different meaning your body may react differently to how other people react which is why you should do testing on a regular basis. Your doctor would be a good starting point, but remember most family doctors are not trained in the latest longevity science and as a result may not take it seriously or disregard it altogether (my doctor certainly did), however the science is profound so you should ensure your doctor is open minded enough to handle new science and advise you accordingly.

It is easy to create charts for visual feedback, so get your results and put them into a chart. You can download an app easily to do this. Compare your chart when you do follow up tests to see how you are progressing. This visual feedback makes tracking your longevity progress easy and will give you a "wow" factor when you see that your efforts are paying off.

Informatics – Longevity Data Tests

Insulin Resistance Tests
Insulin
HbA1C
C-Peptide

DNA Mapping
Genotype Full Spectrum
Klotho
FOXO
Wnt

Thrombotic Risk Tests
Fibrinogen
PT/PTT/INR
D-=Dimer

Glycemic Markers Tests
Hemoglobin A1C
C-Peptide
Insulin Resistance Score
Glucose
Insulin During Fasting

Immune Function Tests
Absolute CD 3 count – T cells
Absolute CD 4 count – helper
Absolute CD 8 count – suppressor
Immune cell counts
CD4/CD8 ratio
% CD 3 Lymph's
% CD 4 Lymph's
% CD 5 Lymph's

Telomerase Measurements
Full Spectrum Mapping

Liver, Kidney, Glucose Function Tests
Uric Acid
Sodium Calcium
Phosphorus
AST
LDH
ALT
Bilirubin
Albumin
Creatinine BUN
eGFR

Liver, Kidney, Glucose Function Tests Continued -
BUN/Creatinine
Glucose
Triglycerides
Potassium
A/G Ratio
GGT
Globulin
Chloride
Alkaline Phosphatase
Protein, Total
LDL Cholesterol
Cholesterol, Total
Iron T. Chol/HDL Ratio
HDL Cholesterol
Estimated CHD Risk

Lipids Profile Full Spectrum Tests
Apolipoprotein B
LDL-C and HDL-C
Insulin Resistance Score
Small LDL particle number
Total Cholesterol
Triglycerides
LDL particle number and size
HDL particle number

Hormone Levels Tests
DHEA-S
Progesterone
Vitamin D
Thyroid (TSH)
Growth Hormone Metabolites
Total Testosterone
Estradiol
Free Testosterone

Prostate Test
Prostate Specific Antigen (men only need apply)

Growth Factors Tests
Insulin Like Growth Factor 1 (IGF-1)
Insulin During Fasting
Insulin Like Growth Factor Binding Protein

Blood Cell Count Tests
WBC Count
RBC Count
Hemoglobin

Blood Cell Count Continued -
Hematocrit
MCV
MCH
MCHC
RDW
Platelet count
Monocytes (% and absolute)
Eos (% and absolute)
Basos (% and absolute)
Neutrophils (% and absolute
Lymphs (% and absolute)
Immature granulocytes
Immature cells

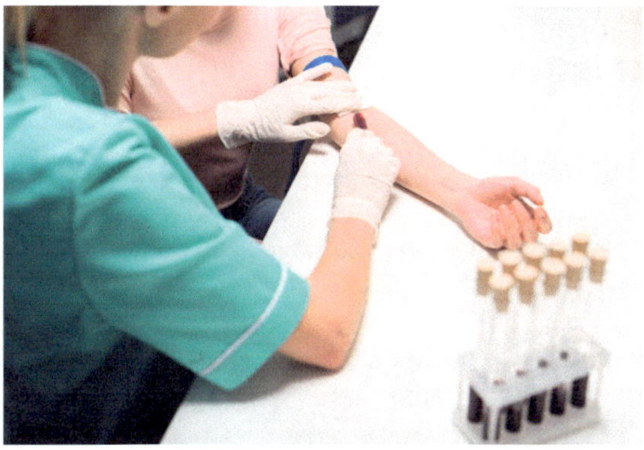

Inflammatory, Heart, Cardiac Tests
Interleukin-6 (IL-6)
C-reactive protein
Fibrinogen
NMR Lipo-profile
Homocysteine
Apoliproprotien D

This might seem like a lot of tests to take but it is really only one visit to your doctor then your local pathology clinic to get the samples taken. It is simply then a case of putting those results into a spreadsheet so you can easily compare later (if you feel the need).

Why are all these tests so important? Here is an example, if you test your immune system for immune senescence this year and find it is working well, but then again next year and notice a severe drop in immune function, you can arrest that drop and take immediate measures to increase your immune cell count and functionality. Longevity is not just about ingesting fancy molecules and staying healthy, it is about data management as well, these tests are your window into how well you are doing and what to take action on. Without that data you are simply flying blind and hoping for the best, and you will learn more about this on the next two pages.

Each country is going to charge differently for these tests and each company that performs those tests will either hand the results to your doctor or to you personally. You want to get those results and maintain a record so that you can watch the biomarkers of aging either drift up or down (note some biomarkers should always stay the same, whereas you may want others to drift up or down depending on your age) so that you can take the necessary action.

The Three Tiers to Biological Escape Velocity

Through my work I have been able to splice extreme longevity into three different tiers, and it is probably wise that I explain those below and on the next page so that you can fully grasp the importance of each one.

Tier 1 Staying healthy, taking pills, exercise, fasting, your diet.

Most of the practical techniques you have read so far such as diet and what pills to take are Tier 1 techniques.

I want you to pretend you are a car, and right now with Tier 1 implemented you are well serviced and the engine is running well.

However, it is night-time, and you are speeding up a dark road with no headlights, you cannot see the dangers in front of you. This can be extremely dangerous to your life (you would never intentionally do it), but this is most likely how you currently cruise through life, playing roulette, hoping your number (a disease diagnosis) does not come up. This is where Tier 2 comes in.

Tier 2 Genetic testing, blood testing, MRI scans, electrocardiogram, etc.

Turning on those headlights with pre-diagnostic genetic testing can not only predict hidden diseases ticking away like time bombs but can even predict rough time of death (to around 3.6 years). This knowledge is confronting to have but can also motivate us to make the changes necessary to live another 50 years. Time of death can be changed with clever therapies and techniques. And pre-emptive disease detection and suppression is also completely possible with today's technology.

I perform this type of longevity service and multiple genetic tests among many other tests (and many of these tests you have probably never heard of). Those tests are company intellectual property, so I cannot elaborate here too much.

If you are serious about longevity, then Tier 2 diagnostics are extremely important for your future. You can garner fragmented Tier 2 data from many other companies, however knowing the interactions and the biological threat reactions between different biosciences and how they affect longevity is the holy grail, and there just does not appear to be anyone offering this facility except my team.

Here is an example. As you already know by now, I sit on the cutting edge of human longevity, passionately researching the drivers of the aging process and devising techniques to combat it. With the tests I have mentioned above, I take all that data, and I cross reference everything. For example, a microbe inside the microbiome may breakdown a particular food and leave a particular type of molecule that activates a disease gene that has been laying dormant inside you. To be able to perform this in-depth analysis takes a particular type of person with a very deep understanding of multiple biosciences, and this is probably why such a service is so rare. It takes a longevity researcher combined with genomics, biochemistry, microbiomics etc. Professional Tier 2 research is an extremely deep dive into multiple biosciences that few scientists care to enter.

And Tier 2 is not cheap due to all the testing that is required to do it properly, however it is certainly within the reach of those that are serious about longevity. How much did you spend on your house or car this year? How much did you spend on your last holiday? I would urge you to consider that your most valuable asset is your health. And that the most important investment you can ever make, is to invest in your health. Without your health, everything else (share portfolio, boat, car, etc) is meaningless.

Our biology is the most sophisticated and complex thing we will ever own and understanding our own unique individual biology should be extremely important to you. Aging is the culmination of a myriad of biological pathways that interact and change one another over time, from genomics, to epigenomics, metabolomics, proteomics, biochemistry, microbiomics, pharmacogenomics, stem cells, and more. Understanding how each of those cause positive or negative effects in another is the crowning glory for Tier 2 research.

Tier 2 reflects the new era in disease prevention. Never have we had the ability to detect cancers, neurological decline or heart disease years before symptoms show. Tier 2 data should be embraced.

Tier 3 Gene Therapies – Personal Engineering

So, we have already discussed gene therapies and CRISPR, and this will certainly become the future as we understand what genes keep us young, and what genes make us old. We do not necessarily need to cut those genes out either, we can use many other techniques, either epigenetic machinery to control gene expression or knocking in (inserting) new youth genes to increase the amount of youth proteins that we know are responsible for various youthful function and rejuvenation. The targets I see are vast, and the technology is already totally possible, none of this is science fiction, it is just a matter of time before we have gene clinics in your local suburb offering this like Botox is offered now. So Tier 3 technologies are coming, and one of my companies is spearheading gene therapy for longevity within the next few years, so please stay tuned, because we want aging to become optional. I am trained in genetic engineering and genetics through Harvard, and I am more than happy to be my own guinea pig in the very near future, because I know I am in safe scientific hands.

You might notice that I write as if aging is completely solvable. That's because I believe it is. My view is no more controversial than Copernicus and Galileo (fellow astronomers of mine) saying that heliocentricity (the sun being at the centre of the solar system) was accurate over geocentricity (Earth being at the centre). Most of the population laughed at those two men, however history is not littered with the sheep that concerned themselves with fitting in, it is sprinkled with visionaries that dared to think differently than the masses and took on what others thought was impossible by following their instincts and passions.

You can read my full paper titled Three Tiers to Biological Escape Velocity – The Quest To Outwit Aging at www.pubmed.ncbi.nlm.nih.gov/36606268/

Hit the view PDF button on that page to obtain the full transcript (which is completely free).

Quick Guide – Step 2 The Toolkit

Not interested in reading the science? Are you in a big rush to live forever? That's fine, but only after speaking to your doctor first, should you take these supplements and implement this regime.

First and foremost, do some blood tests to establish a baseline. Secondly, diet and exercise. We have already covered the type of food you should lean toward and we have covered the best exercise techniques, and later we cover some simple exercises. The supplements listed below all have their importance along the aging pathways and I am a user of all of them. However, your budget may prevent you from using them all.

To determine dosage, use the instructions on the bottle from the company you purchase from and speak with your doctor.

- Metformin – Reduces your chances of disease (depending on the condition). 1500mg to 2000mg daily is recommended and must be used under your doctor's supervision. One 500mg tablet every four hours is often advised, but again only your doctor can advise you properly. Take a vitamin B12 supplement during metformin use.

- Rapamycin – Appears to increase lifespan and resists muscle wasting among other benefits.

- NAD – Maintains a cell's energy levels, prevents vascular aging, ensures proper cellular function, and reduces cancer risk.

- Spermidine – Autophagy, telomere length, decreased arterial aging.

- Urolithin A - Mitophagy

- Collagen – Skin, bones, ligaments, cartilage,

- Vitamin K2 (Mk-7) – Arterial health, reduced calcium deposits, osteoporosis.

- Sirt 6 Activation – Age delay, youthful metabolic activity, elevated NAD

- AKG – Cellular energy, increased metabolic capacity.

- Creatine – Stores energy inside your muscles ready for heavy activity, giving you more youthful function.

- Curcumin – Switches cancer suppressing genes on (epigenetic diet), incredibly powerful anti-inflammatory.

- Pterostilbene – Extremely powerful antioxidant.

- Resveratrol – Turns on splicing factor genes and reinvigorates senescent cells, activates the Sirtuin genes and more.

- Melatonin – Closes the mitochondrial permeability transition pore (among other things), natural hormone which also is a powerful antioxidant.

- Quercetin – Scavenges free radicals, prevents gene mutations.
- PQQ – Creates new mitochondria.
- Fisetin – Purges senescent cells.
- Dasatinib + Quercetin (take together) – Purges senescent cells. 2.5mg per kilogram of body weight for dasatinib and 25mg per body weight for quercetin. Ensure you speak with your doctor first. Note this use of quercetin is different than above.
- AMPK – Enables perfect signaling between cells and cellular machinery.
- Astaxanthin – Mitochondrial protector among many other things.
- Acetyl L Carnitine – Anti-glycation, mitochondrial protection, take with CoQ10.
- Ubiquinol aka CoQ10 – Assists with the production of adenosine triphosphate (among other things).
- L-Arginine – Slows or prevents vascular aging.
- L-Carnosine – Reinvigorates cells, reverses mitochondrial dysfunction, anti-cancer
- Glutathione via NAC and Glycine – One of the body's most powerful antioxidants
- Hyaluronic Acid – Structural support, lubricator, and hydration. Medium for cell transport.
- Diet – Eat non-processed, low sugar, non-animals-based foods and keeping as close to nature as possible extends life.
- Control epigenetic drift – Increase gene expression with your epigenetic diet.
- Control the formation of Advanced Glycation End Products with our good friends Carnosine and Pyridoxamine, Acetyl L Carnitine and a low sugar plant-based diet (or GlyGator).
- Fasting – Purges dysfunctional cells, releases new stem cells and builds brand new tissue, lowers blood sugars, slows down the formation of AGEs and much more.
- Exercise – Perform three or four different types per week. Weight training two days per week, fast walking the next day, aerobic exercise (running, boxing, tennis) the next day, etc.
- Meditation - Reduces inflammation, cortisol, cell-mediated immunity and biological aging and is shown to influence gene expression.

Only quick explanations are given on this page, each product above does many more things, however read the sections for each supplement to learn more.

Ray's Regime – Or Am I Just A Pill Popping Preacher?

I wake up and do three minutes of mountain climbers (three sixty second bursts with a thirty second break, total four minutes on the ground). This high intensity training immediately kicks in metabolic, biological and longevity pathways (such as NRF2) that will lower my blood pressure and put me into fat burning mode, especially because I have not eaten since 6pm the night before (and I will not eat for at least another four hours). I fast from 6pm (sometimes 7pm) till lunchtime the next day. The craziest thing is, the less you eat, the less hungry you feel. I may have a green tea or soda water in the evenings. I'm no saint though, because I am lean and fit, I can easily eat in the evening and consume great wine without worrying. My metabolism is supercharged from my fasting, exercise and the pills I take which raise my cellular respiration (which to me, and probably to others that know me, appears like a twenty-year-old).

Remember, exercise creates a huge number of free radicals; it is your body's response to the exercise that purges these free radicals and also performs a myriad of other facets that kick in the longevity magic.

I will make a coffee which includes a green tea bag and a about a gram of curcumin (I first mix the curcumin with coconut oil otherwise it will not dissolve). You will need to engineer this type of drink to suit your taste. We (my cohorts of other longevity experts) have nicknamed this drink a Griffin (GReen tea, coFFee, curcumIN)

The reason Griffins are handy, is because you can get all the benefits from three different types of longevity food groups with one drink. I have about two or three per day. I also drink earl grey tea with a chamomile tea bag simultaneously. The chamomile contains apigenin, which is great for your anti-aging toolkit. The earl grey is simply for the polyphenols (other dark teas are just as good). I am yet to name this combo.

I'll take pills throughout the entire day (mostly with food). I take 500mg of metformin, but not within 6 hours of training (either side), and 500millgrams of nicotinamide mononucleotide (NMN to raise my NAD levels of course), PQQ, Collagen, 5 grams of Creatine, 10 grams of hydrolyzed collagen, Urolithin A, Vitamin K2 (Mk-7), Cyanidin or Oleic Acid (Sirt6 activation), DHEA, an AMPK regulator, Astaxanthin, Co-Enzyme 10, AKG, 60mg of Hyaluronic Acid, NAC, Berberine and TMG some day's alpha lipoic acid, Folate, B12, Carnosine and 3mg of Melatonin at bedtime (and a few other sprinkles of magic here and there). I also intermittently take combinations to purge senescent cells, prevent glycation, raise my klotho levels, extend telomeres and (very important) prevent further epigenetic drift. I no longer take spermidine, quercetin, resveratrol or pterostilbene as the science is conflicted and points to minimal benefit (if any).

I have a tablespoon of apple cider vinegar around lunch time, and then again later in the evening.

I'll add another two minutes of exercise a day, usually I do thirty chin-ups (palms front) in one go. Sometimes I may do thirty jump squats (these all take less than one minute). You can spread these exercises throughout the day as well. You are literally doing two or three minutes per day, and if you do them close together, because they are high intensity, you are getting the same benefits from someone who trains for 45 minutes per day. Personal trainers may argue with that, but the science doesn't. Checkout Michael Mosley's The Truth About Exercise (online). These short sudden bursts of energy keep my NRF2 pathway activated and hormesis active.

However, with exercise you might have back problems, you may be older than me, you may have heart trouble, speak to your medical professional about what type of high intensity training could be good for you. A seventy-year-old may not be able to jump squat but could hold something to stabilize (a door frame maybe) and squat up and down twenty times, keeping the back and neck straight. Everyone is different, so I can't offer tailored advice here, but any type of muscle building and elevated heart level can activate longevity pathways, not to mention help extend telomeres (and of course, burn fat). The exercise also reduces glucose floating around your blood stream, which means it maximizes the fasting effect even further, kicking in those longevity pathways.

At lunch time (when I break my fast) I'll eat something which contains very little energy. Spinach shakes with two strawberries. No foods with high levels of sugar (glucose), carbs or fat, such as bread, pasta, rice, potato, breakfast cereals (among other foods) will rarely be consumed by me. That is because I want to keep the fasting effect, I like the fact that my body is burning fat 24/7 and I am in an AMPK activated state with an mTOR deactivated state. I enjoy the mental clarity it brings and having a flat stomach. If you struggle with weight, I've included a weight loss section at the back of this book (including some food ideas and training techniques).

If I were to order a burger or pizza, there would be a perfectly round bread bun left on my plate afterwards, or a perfectly round pizza base left in the box. I would eat the protein, the veggies, the cheese, but rarely the bread.

I also experiment on myself with a few other things. I use nasal sprays to raise my NAD levels and I'm currently designing a skin patch for myself (and my readers) that can deliver 500mg of NMN per day through the skin.

My diet is pretty lean. I used to eat junk food, I used to eat breakfast, I used to eat late at night (snacking) when I was younger and oblivious to the science of losing weight (and longevity), but being a researcher and seeing the benefits, it becomes hard to ignore the data you are seeing. Now I really do find eating lean and being fit much more rewarding that having a quick fix of some sugary junk food. How many of you look at yourself in the mirror and wish you could remove some troublesome fat? Read the weight loss section at the back of this book and checkout some longevity foods I have also included at the back. I'm no chef, but this book will hopefully give you an idea on when and what to eat. I've also rated those foods with a longevity scale.

Maybe two days per week I'll ride my bike around the block. Each time I ride, I try and beat my last time. Remember, high intensity interval training will put you into longevity mode. We are not talking about a Sunday stroll or relaxing bike ride to look at the sunset. When I finish my ten-minute bike ride, I feel like I am dying (metaphorically), I am out of breath, I am soaked in sweat. This was only ten minutes and is something to aim for.

I stand at a computer (sometimes a microscope) for most of the day and try and conduct meetings standing up as well. I rarely sit. I always take stairs instead of elevators or escalators. I'm always surprised that after sitting on a plane for five hours, people still take the elevators down one level to baggage claim. Learn to move so that sedentariness becomes like poor hygiene or poor English; it just starts to feel wrong at a subconscious level. Any phone calls I get during the day will see me pacing in circles for however long some of my phone conferences go for. You get the picture.

Use the one percent rule. Start making small changes with exercise, diet and fasting. Slowly ween yourself away from sitting when you get a phone call. Take the stairs not the escalators. Start having smaller breakfasts until you are not eating breakfast at all. Etc. Let this information sink in over time. Get a highlighter and mark sections of this book that you feel are most relevant to you as well. Books like this which are packed with data can be hard to retain, so keep it handy and revert to it regularly.

I also perform four thirty-minute weightlifting sessions per week. Combine this with my short high intensity exercises and maybe a quick ten-minute bike ride, and my exercise routine is around three hours per week in total.

I am no elite sergeant major in the marines, if I am tired or busy, then my training is just one-minute bursts of various exercises throughout the day (pushups, chin-ups, tricep-dips, jump squats, etc).

A lot of people are going to tell you that you cannot get healthy with small bursts of high intensity training. I again urge you to watch Michael Mosley's "The Truth About Exercise".

I take cold showers as well. I take myself out of the thermo-neutral zone by having a cold shower or sleeping without so many layers on. You can get the same effect from wearing fewer layers in winter.

And science also now shows us that a 45-minute hot bath can activate heat shock proteins that can assist with preventing sugar spikes from high glucose foods. So, I take at least one hot spa a week, more if time permits. Remember that adaptive hormesis is your friend.

I throw in some meditation, sleep and in the extremely near future some CRISPR to knock out those pesky aging genes to ensure I age well.

This is my routine. It is not a diet or a training camp, it is a lifestyle that slowly took shape over a few years (certainly not overnight) and now comes easily. I was once carrying excess weight, eating improperly, and making all the other mistakes people make. However, I have honed the way I age and you can to. It is a great feeling being in control, looking and feeling good.

And a final note, I have introduced you to about thirty different oral supplements in this book. I take most of them, but a few of the lower scoring ones I don't. I feel I already do enough to resist aging, so aim for the higher scoring products if you are on a budget or don't feel comfortable taking so many.

The Scientific Method

Science is not about giving us a sense of security; nor is it about making us feel warm and fuzzy; and it is especially not about having interpretations of natural phenomena simply because they sound good or make us feel good. Science is about where the data takes us regardless of what you believe, regardless of what makes you feel good and regardless of what you wanted or expected. It is a discipline that once learned will deliver a higher state of being.

The world is full of pseudo-scientific people that will warp science into their own version of science as they try and make it fit into their own perceptions. You must be cautious that your lack of scientific expertise is not exploited by charlatans. This is one of the reasons that at the front of this book I tell you how to do your own research on .gov and .edu sites as opposed to .com and .net sites.

The scientific method is about observing, questioning, establishing a hypothesis, running experiments, and performing research, analyzing the results, ensuring the results are accurate, then forming a conclusion on how it happened. At which point, you share the data with other institutions or groups and if they can also replicate the same results, then you have science. However if they cannot replicate the results, you do not have anything scientific but merely a random result (a claim, a statement), that could have come from anywhere.

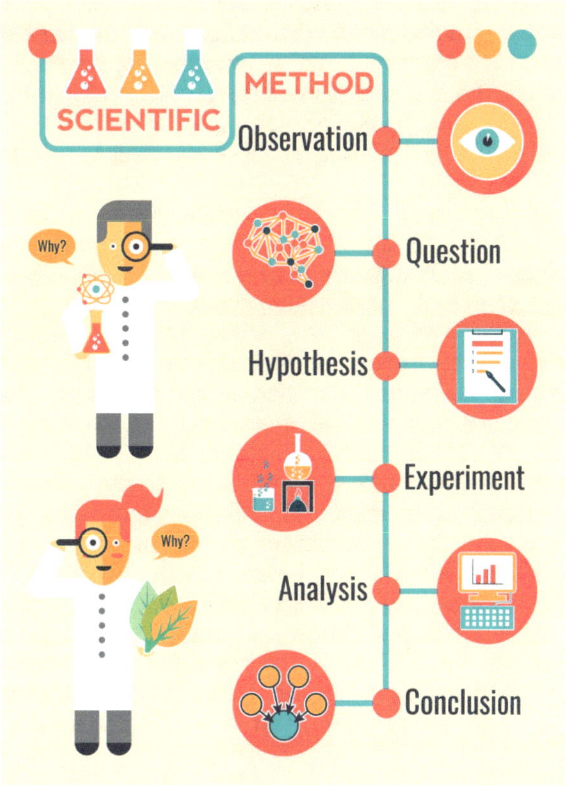

How to spot a charlatan? Anyone claiming to be able to cure or treat any type of disease or condition should **first have** peer reviewed clinical trial data and results prior to making such claims. This is the biggest red flag. I have spoken to employees of companies selling anti-aging / longevity technologies and as soon as clinical trial data or proof of concept was asked for they run a million miles. This information is important to your anti-aging toolkit to ensure you are not wasting time and money on fads that sound great but simply do not work.

You have probably noticed while reading this book that the technology I introduce you to also comes with names of the scientists or the institutions that founded them. This is not just to show respect to those pioneering scientists that are bringing extreme longevity closer, but it is also part of the scientific method, so that you, the reader are aware that these are not just random claims or beliefs. Maintain that standard across all the literature you read. In fairness a good idea can still be pitched and may work without clinical trials but claims of cures must be backed up with rigorous data.

Raymond Palmer's Triple R Theory on Aging - Replacement, Rejuvenate, Repair.

This is my considerably basic version for your toolkit, and not a deep dive into my science.

So, what's my theory on ageing? Do I have one? Of course, and it is a relatively effective approach. I have found that there are three categories to the aging process. Things we lose (replace them), things that deteriorate (rejuvenate them) and bad things that accumulate (remove them). It goes something like this –

Replacement. Everything that tapers off during age such as hormones, NAD, the body's antioxidants like glutathione, the body's enzymes, cell function should be replaced with clever compounds, enzymes, co-enzymes etcetera to restore biological pathways and function to a youthful state. That is to replace and restore all levels of biochemistry to youthful levels. Anything that tapers off after youth should be restored. Including the epigenome, in 2019 Dr Greg Fahy came across a way to wind back the epigenetic clock (great work Greg), so a youthful epigenetic state is also now replaceable.

|

Rejuvenation - regrow and restore muscles, blood vessels and other tissues with stem cells, enzymes, steroids, exercise, human growth hormone, NAD and other wonder drugs or compounds. In fact the steroids such as testosterone or human growth hormone could also be considered for the first category (replacement), there is certainly overlap between the Replacement and Rejuvenation categories. Replacement leads to rejuvenation, so think of rejuvenation as being downstream of replacement. Telomeres also need rejuvenating along with the thymus gland and many other organs. Speak to your Doctor, but do not expect a longevity expert either, as Doctors are not trained in longevity, but trained more to diagnose and treat illness, disease and injury.

Removal - removing the damage inside the cells, such as mutations in mitochondrial DNA and Nuclear DNA. This is trickier than it sounds. However, we are working hard on ways for restoring genetic integrity. Other wasteful junk inside the cells also needs removal, and there are ways to boost these processes that I have written about already. We also need to factor in the removal of hormones or enzymes that can cause aging or damage. For example, eNAMPT is an NAD guzzler that rips away at your NAD levels as you age. We want to remove or slow down anything that causes damage, not just the damage. Drugs like metformin and fasting can also slow down cell division, thus resulting in less replicative errors. So here we are removing the cause to that damage, and not just the damage.

So, it's that simple people (maybe). Aging seems to have three categories. Treatments for the first two categories, replacement and rejuvenation are already here. The third category, repair, especially of genomic integrity is a harder target. However, keeping your genes undamaged in the first place is your first important step (no toxins, healthy lifestyle will certainly help). CRISPR gene editing will certainly play a big role in correcting damage in future years, we will be able to sequence a person's genes and hopefully find dangerous mutations which can then be knocked out with gene editing. So treatments for this are already here and just need to go mainstream. The tools have now landed, learning how to use them to solve various genomic problems is our next step. When I think about the theory, we really are not that far off from biological escape velocity.

Ray's Future Projects and Research

I have recently started up a new company which will be conducting the world's first human trial in tissue rejuvenation. I have already developed and patented the technology, and now it is just a matter of testing it. So, this human trial which we are pushing to have done within twelve months should be able to restore youthful function back into tissues such as brain, heart, lung, liver, kidneys, skin, bone and muscle. Yes, one therapy, one patented pharmaceutical may do all that. Even if my technology only does 30% of what I say it can, the rewards for humanity are vast and the first real therapy against aging is officially on.

This technology will not only rejuvenate the above tissues, and maybe more, but will hopefully prevent an insurmountable amount of disease and possibly even reverse some diseases or conditions found in those tissues. This is because many diseases of old age are caused by tissue dysfunction, so keeping those tissues young and vibrant may restore function and cure disease.

Personally, I can't wait to show you how my therapy works, it's quite elegant, and hopefully extremely powerful. At this early stage, I'm legally bound to keep things quiet, but once the human trial is on, you probably won't be able to miss it, because it will be the first trial that aims to reverse aging.

Crackpots versus Scientists

When Galileo first said the Earth orbited the sun, he was met with disdain and abuse, by people I might add, that had zero scientific training and had not even tried to examine his observations. Humans have not evolved much past this, because when you work in a fringe science like me, you get to see how resistant people are to changing paradigms. I have been abused by simply telling people I work on human age reversal technologies. I receive regular hate mail from environmentalists, religious fundamentalists / whack jobs and anti-vaxxers. Apparently, scientists are the devil. Some of the emails are fantastically entertaining, with one lady telling me I need to perform research on the healing energy of crystal chanting (whatever that means) to other emails telling me I clearly work for Bill Gates and must be a pivotal piece to the de-population conspiracy theory to kill millions of people (the list of crazy emails I get goes on). You don't have to look hard to find me getting abused online.

All scientists that made breakthroughs somewhere in history were attacked. A Nobel winning scientist here in Western Australia, Dr. Barry Marshall was attacked by other scientists when he said that stomach ulcers were caused by bacteria in the gut. Fellow scientists told him he was a crackpot because bacteria could not exist in the gut. The good Doctor was shunned professionally, so he drank some bacteria (Pylori), and gave himself a stomach ulcer, which he then cured with antibiotics, which led to a Nobel Prize.

Interestingly, not only can bacteria live inside the gut, but we now know, we cannot live without bacteria in the gut. Paradigm change is very difficult for people to accept sometimes, even some scientists.

Fulfilment?

There are many who believe that because we are all destined to die with an average life span of about 70 years, that this finite time is what forces us up out of bed each day to achieve things because our time is finite. Fair enough, you cannot really lounge around for 20 or 30 years if you want a family or education, you need to move on these things early in life when your average life span is 70.

Nevertheless, there are others that say the opposite, that if you have unlimited time or much more time than 70 years you will be overwhelmed with choice due to the amount of things you can do. Learn a few musical instruments, learn to paint, travel to every corner of the globe, learn a few languages, choice will become incredible for those that achieve extreme longevity.

And if you are capable of living a few hundred years or more, how long are you supposed to stay with the same person for? What does "till death do us part" mean for marriage to those that have achieved extreme longevity? No longer needing to reproduce and raise children may also see relationships as quite pointless for future generations. Or maybe moving from partner to a new partner every few years will become the new norm with marriage vows such as "till boredom do us part".

Extreme longevity and the global effects is still a path unknown to humanity. With medical breakthroughs such as vaccines and cures for diseases, healthier living and longevity technologies, the life expectancy is set to explode further. It is like the human population may be on a collision course for being scientifically too successful.

Many people will say that we should be destined to die as nature programmed. Well those people should be the first to cease visiting hospitals or doctors whenever nature programmed them to get sick. All forms of healthcare are a form of longevity intervention, so where do we draw the line?

Let's not kid ourselves, there is no such thing as 'aging gracefully' waiting for the onset of degenerative diseases such as arthritis or Alzheimer's. Ask anyone with these diseases if they found them 'graceful' and you certainly will not hear them say nice things about the experience. Using the word graceful is almost an insult when dealing with human suffering. The experience can be like a nightmare for those that are unlucky enough to experience some age-related disease. Any technology that reduces human suffering and improves our quality of life has a place in a civilized society. Those that would deny a modern society of such technology are welcome to not use it. And those that choose to not use it for whatever reason are sleep walking into death.

"Do not go gentle into that good night,
rage against the dying of the light"
(words from Welsh Poet Dylan Thomas 1945).

Society and the World

If our societies stop aging what will that mean for our population? Will we have 10 billion people walking around that are no longer aging or dying? If so what does that mean for our requirement to reproduce and have children? Would there be an over population of old people burdening us? Not likely. Many modern sophisticated countries such as Japan, Italy, Monaco have negative birth rates and the whole idea of longevity is to keep you young and vibrant so that you can continue to be productive and not burden the healthcare system. That silver tsunami of old people that comes to mind when we discuss extreme longevity is a myth, the technologies in this book are designed to keep you young, not just extend life.

It is no secret that countries with rampant breeding and over population have increased levels of poverty (and in some cases famine), so keeping millions of humans young and productive to assist in ensuring we have a vibrant society full of experienced people that have already lived for many decades could help prevent the problems created by inexperienced populations.

It is only when humans are not kept young and vibrant and over population becomes a burden with unsustainable elderly people burdening struggling societies and infrastructure that we could see governments begin to regulate the population by selecting who is allowed to have children? Who is allowed to stay young (those that can afford it)? Those that can financially support their children or maybe only those that are better educated may be able to have children? These are concerning arguments that could possibly be made in a world where over population has become a major issue. Keeping people young, healthy and contributing would certainly see those risks become minimized.

And let's not pretend the world can hold an infinite number of humans, it simply cannot and population growth is exponential in many countries, so eventually responsible breeding and worldwide population control is probably closer than you think.

But could this population control result in longevity technologies being banned to keep the population down? Forcing older people to die as nature intended? Probably not, seeing as keeping people young and productive means populations are able to cost less and produce more, so the inverse of forcing people to die would likely to be true.

And will these radical techniques for immortality or extreme longevity be cheap enough for the entire planet to use? Or will they be reserved for just those that can afford it? Again, another facet to extreme longevity that would be quite concerning for many people, however most of the technologies shown in this book are already cost effective and with the more expensive radical therapies are becoming mainstream, so those too will see cost effective pricing.

When we discuss the price of longevity, where do you draw the line between longevity therapies and Medicare therapies? Both types of therapies are designed to keep you healthy. But here is the dichotomy.

- Healthcare systems are not interested in keeping you young or healthy. They only attempt to fix you once you get a disease. This is expensive and highly risky.

- Longevity technologies attempt to keep you young and prevent you from getting disease in the first place.

So really, where should your healthcare system be putting your tax payer dollars and efforts in the first place?

And if we can save billions for our governments by not getting sick and frail in the first place, would it not make sense for our governments to start supplementing longevity technologies for their populations? Well governments already do supplement technologies that prevent you from getting sick such as vaccines. Longevity technologies are just the next step in this phase.

For me, this is common sense and obvious, but history shows us that some world leaders are extremely slow in the uptake of new ways of thinking or technology. Galileo was a great example of that when he tried to show the Vatican (the government at the time) that the Earth orbited the sun as opposed to the sun orbiting the Earth. He was laughed at and sentenced to death which was later repealed to home detention for life. The point is that he was correct whilst the masses and world leaders were of course wrong, they were simply not open to new ideas. Humans are still averse to new concepts that they are not accustomed to as a child (any deviation from their reality is difficult for an indoctrinated mindset to accept). As an inventor, astronomer, researcher of longevity technologies and biomedical engineer, I live in an environment where new ideas are a pre-requisite for my work, so I see the human condition more than most when it comes to narrow minded people, I can pick them in an instant and choose very carefully who to discuss my work with when out on social engagements. As an example, just try and explain some of the longevity technologies in this book at your next family dinner and watch the smirks and negative comments fly. The interesting thing is that those that do smirk have not spent one minute researching the field, so feel sorry for their narrow mind that they feel empowered enough to judge a field they know nothing about – which is the absolute epitome of a closed mindset. It demonstrates an indoctrinated brain, a pre-programmed mind to auto-respond to any stimuli that may breach their paradigm (they are plugged into the Matrix). Stay your course in the face of lost souls, you are awake, they are not, Galileo and many others had it much harder.

Intelligent population and breeding policies will come. The requirement to get off this planet is also coming. Longevity technologies will also be paramount for space and especially interstellar travel. So expect major developments and discoveries over the next two decades, and hopefully you can live long enough to implement them.

"Though wise men at their end know dark is right,
Because their words had forked no lightning,
they do not go gentle into that good night"
(second verse, Dylan Thomas 1945)

Dylan Thomas wrote that poem for his father, who after a life of hard work, acquiring wisdom, knowledge and wealth would then be rewarded with death. Thomas was obviously frustrated with this deal and maybe you should be too.

Over The Last Few Years

The reason we were originally programmed to die at an early age (30) was because in early evolution food was scarce, so our species died young to ensure the survival of our species. Fast forward a few million years, and our species has now developed dozens of ways to keep us alive with vaccines, antibiotics, diet and farming along with many other systems. It is no longer imperative to die young to prevent us from extinction. The alternative may be true; longer lived people with more life experience may actually prevent the species from killing itself.

Over the last few years humans, have developed some of the biggest technologically advanced systems to keep us alive such as –

- CRISPR-CAS 9 (gene editing)
- CRISPR-CAS 12a (gene editing)
- Stem Cell Rejuvenation
- Immunotherapy (curing cancer with tailor made immune cells for each person)
- Artificial Intelligence which designs drugs (compounds)
- Cure Hepatitis C

These are now billion dollar industries that did not even exist a few years ago. Millions of people remain alive or will remain alive because of these very recent technologies. If you are paying attention to the trajectory curve here and watching how fast science is moving at keeping us alive and young, you must be aware that longevity escape velocity is going to be here in the blink of an eye in evolutionary terms.

It would be completely rational to predict that the curve for longevity technologies that specifically target the aging pathways will be even bigger and more popular, because what person does not want to stay young and healthy? The entire world will subscribe. But what about those that are resistant to staying young and healthy? If someone does actually claim they want to become old and sick then tell them that's okay (it is their choice), but as long as they can do it without burdening our healthcare systems and pay their own way through, then ethically who has the right to stop them? The ethics of the very near future are of course keeping people young and healthy. But it is an interesting vision of the future that the pressure will not be about overpopulation, but be on those burdening our healthcare systems by being too lazy or too stubborn to buy affordable longevity products to shut down the pathways of aging. So non-evolving people beware, we are coming.

Message From Me – Raymond D Palmer

At this point your brain may be spinning. You have read a mountain of information. So, I am going to say a quick goodbye here.

However, if you need help losing weight, knowing what foods to eat or how to do quick high intensity exercise, I have placed three sections after this page. If you already have these things under control, you don't really need to read them, but it may pay to take a quick look and scan the pages for intel you may not know.

I am currently putting a team together and building a biotech company in Dubai, and we will run the world's first human trial in human tissue age reversal. When I say tissues, I am talking about the brain, skin, muscles, kidneys, liver, lungs, heart and much more. You will most likely see this on the news somewhere, as the clinical trial documentation requires us to name the disease we are treating, and the disease is of course "human tissue aging", so no doubt some keen-eyed journalist that monitors these things may start asking questions.

Now that you have read the bulk of longevity science, whichever type of anti-aging architecture you decide to use, the amalgamation of various technologies in the coming years will either resonate like a symphony of harmonious biochemistry or squeal like a twisted tangle of abhorrent side effects. I know human nature well enough to know that some readers will chase anything I write here, so I'm careful with how much and what technology I've shared. I will not mention many of the other studies and technologies yet because they are quite risky and the jury is still out on their safety (but there are many and they are vast).

Experience shows us that society is full of risk takers and therefore many other enzymes or compounds and techniques to prolong life have not been included. Some readers' intent on the goal of immortality will tend to try anything and I will not arm them with too much information unless we have some trustworthy science behind it. But alas, Moore's Law that was first coined in a 1965 paper about integrated circuits tells us that technological ability doubles every two years. Decades later and Moore's Law still holds true. So if we double the technology that is in this book, Type II and Type I status may not be far away at all.

Hopefully this book has given you a great foundation into healthy aging and prolonging your life. I am working hard to establish the ultimate anti-aging regime which includes therapies from the radical longevity section so that we can control all known aging pathways, not just one or two pathways, so follow me and keep in touch on Twitter, Facebook, Instagram, YouTube, and scan the QR code at the back of the book and stay tuned because you've seen nothing yet.

Did I miss something? Have you read a conflicting study? Want me to cover some other technology or anti-aging science in the next edition?

Are you a distributor of anti-aging technology? Want me to review your science? Then contact us at the above, we are open to moving new technology forward for consumers but only when it works.

Thank you for reading my research review, there is much more to come in future editions.

The next sections are losing weight, some food ideas and exercise ideas to assist you with your anti-aging regime. So note, I separated the following sections and placed them into the back of this book, so that those of you who are already across those facets do not have to read them. I'll repeat that my job in this book is to get you in and out as quickly as possible, there is so much information to absorb that I am very mindful of your time and retention. This book could have easily been 500 pages, but that would have been quite annoying for anyone reading it.

YouTube

There is also a warning on the next page if you are thinking about biohacking. You need to appreciate the risks.

Reach out and say hello! Would love to hear what you think of my book!

Best regards and remember, everything in moderation, except longevity.

Raymond D Palmer – PhD Candidate / Inventor / Immortalist / Author / Astronomer / Biotech Engineer / Futurist / Polymath / CSO / Tissue Reprogramming

As disease tries to steal your light,
Rage against the dying of your life.

www.ScienceOfAging.life

And some eye candy for you -

Biohacking – A Word of Warning

Are you tinkering with the idea of using some of the more fringe technologies you have read about in this book? Read something elsewhere that promises the fountain of youth? They sure sound promising, and some just might be promising. However, lets sober up for a minute.

Biohackers are a growing community of people that are taking serious risks by using unproven technologies. Let me introduce you to Aaron Traywick the CEO of Ascendance Biomedical who provided an unproven AIDS cure to patients.

Traywick was not a conman trying to trick people, he really believed in his therapies. Traywick created headlines himself in February 2018 when he injected himself with an untested Herpes cure live in front of an audience. A few months later, Traywick was found dead inside a flotation tank. At the time of writing his cause of death remains unknown.

Prior to Traywick's death, he had also assisted a 28 year old AIDS patient Tristan Roberts by providing Roberts with his unproven AIDS treatment. Roberts injected himself live on Facebook and stopped taking his conventional medicine. About three weeks after this injection Robert's back developed large reddish lumps and he was also down for a number of days with strange fever type symptoms along with gastrointestinal problems and loss of appetite.

Prior to this injection and while he was still on conventional treatments, Roberts had his blood tested and his viral load count was 11,912 copies of the virus per milliliter. Roberts was hoping to see his viral load drop down with his biohacking. Unfortunately a few weeks later for this biohacker, his viral load had increased to 36,000 viral copies per milliliter.

The point of this page is that Type II treatments are still unproven and could be extremely dangerous. So be careful if you are ever deciding to seek out a type II treatment and make sure you chat with your healthcare professional and get those tests done so that you can monitor your status. Not all type II's are made equal.

Biohackers

Make no mistake, the Longevity industry may become the next big black market across the world with unscrupulous people selling and dealing in fake stem cells, selling fake parabiosis potions, fake stimulating enzymes, etc; and this is all going to happen over the next two decades. How much would you pay for an extra year of life? How much would you pay for an extra decade of life? How much would an 80 year old pay to be 30 again? This industry and the technologies I've written about in this book will be exploited at every level and at every turn. Anyone that cannot see this industry being exploited has not seen deeply into human nature at all. We are born with the ultimate desire to survive and live well. So can you imagine the black market this technology could create?

"Psst, mate, wanna live a few decades longer?
Come down this dark alley"

Take care in what you purchase and how you use it.

Investigate if your suppliers are using the scientific method.

Always chat with your Doctor or healthcare professional.

Weight Loss Section – Understanding the Science

This book is your weapon to weight control. This section is only for those of you that want or need to lose weight. I am not going to bore you with 500 hundred pages of studies and weight loss diatribe. Instead I'm going to take you directly into the nuts and bolts of weight loss (the hormonal science) and what you need to do to put your body in weight loss mode (fat burning mode) and then how to live life well and keep that weight off.

- It kind of amazes me that weight loss companies do not teach the science (biochemistry) of weight loss.

Using simple tricks, I will teach you how to lose weight and keep it off whilst enjoying the life you want.

Empower yourself with knowledge to be the person you want and live the life you want.

How To Use This Section

If you are serious about losing weight or keeping it off then read this section multiple times until the information is clear and easy to remember. Use a highlighter and sticky tabs so you can revert to sections that are pertinent to you. I won't get offended if my book looks like a school book, with bookmarks and scribble all over it.

Even though I contemplated going into much more scientific detail and referencing all of the scientific papers and journals for you to read, I have intentionally not done this. However you may notice a small number of them through this section (reading them is not required).

Ultimately, at the end of the day you are probably reading this section because you would like to learn more about how the human body stores and burns fat.

Claims Not Made

This section is not intended to teach you about health foods; it will teach you how to lose weight! Some healthy foods can take you out of ketosis mode (explanation later in this section), so speak to your doctor if you have any questions. But understand that some healthy foods are not necessarily good for losing weight, and that some foods that are great for losing weight may not be packed with nutrients.

I am not your doctor. And your doctor probably did his or her degree many years ago, and may not have studied this type of science at all. Keep that in mind.

I had a substitute doctor a few years ago while my usual doctor was away, and the substitute doctor told me that white rice was healthy and good for losing weight. I was quite stunned. In his defense, this is not something he would have studied in his degree, and his comment proved that fact.

- **Forget diets, this section will become your lifestyle.**

Controlling Two Hormones – The Biochemistry of Losing Weight

It is a funny thing losing weight. It comes down to two hormones, insulin and glucagon. Insulin regulates the absorption of sugars **by telling the body to lock away the calories you eat and store them as fat**.

Glucagon (your friend) does the opposite, **it tells your body to use your stored fat as your energy source**.

So losing weight comes down to controlling these two hormones with really simple techniques.

However, even though the philosophy behind this techniques is very easy to understand, after a life of eating whatever you want whenever you want you may have to learn some new techniques, with the biggest challenge being self-control.

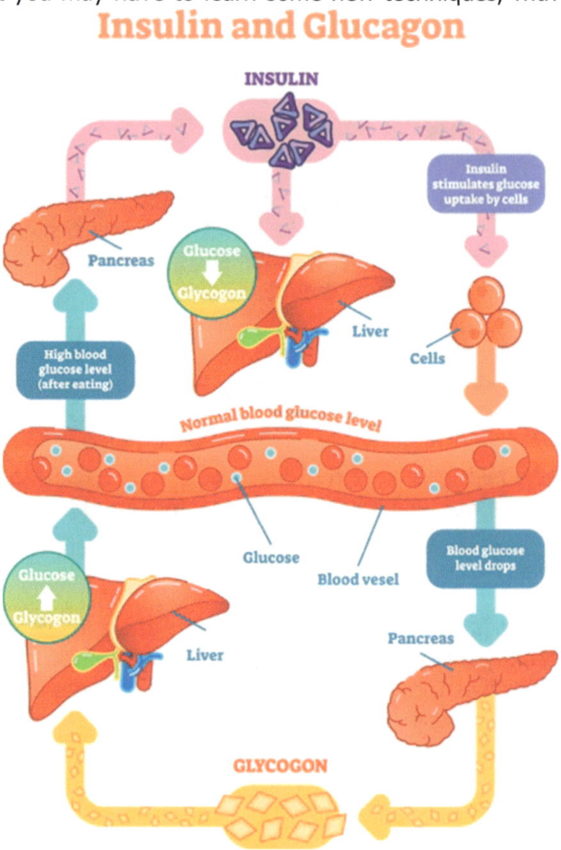

But don't let that term "self-control" put you off from losing weight. Self-control is nothing more than a neural network in your brain that you can build. If you have never had to learn self-control before you may find it difficult at first, but neural networks are like muscles and will strengthen over time. It doesn't take long at all before you find yourself in complete control of your weight and appearance destiny.

So when you get the urge to snack or splurge, simply ask yourself first, how much do you want to look and feel good? Remember how you feel standing in the mirror. The journey starts now, taking control and feeling great now begins.

Comfort food and snacking is your enemy, so learn to view it as such!

The Battle Between Good and Evil - You want the glucagon hormone in your blood to be elevated and you want your insulin levels low.

- **It is only when your insulin levels rise that your body stops burning fat and you begin <u>locking away food that you have just eaten as fat.</u>**

So this is a battle between keeping your glucagon levels high and your insulin levels low. This section will teach you how to do this.

Every time you eat or snack between meals your insulin levels will rise and then slowly come back down. The more sugary your meal is, the higher and longer your insulin levels will stay high (keeping you in fat storage mode, and out of fat burning mode).

Now think about this. You have just eaten a burger with enough calories to keep our ancestors going for a week. Your insulin levels have jumped and your body is no longer burning fat for energy but instead, storing it. You then decide to wash this meal down with a sugary soft drink, spiking your insulin levels even higher and longer which in turn tells your body for much longer to store your meal (including the sugary soft drink) as fat, as opposed to burning the stored fat.

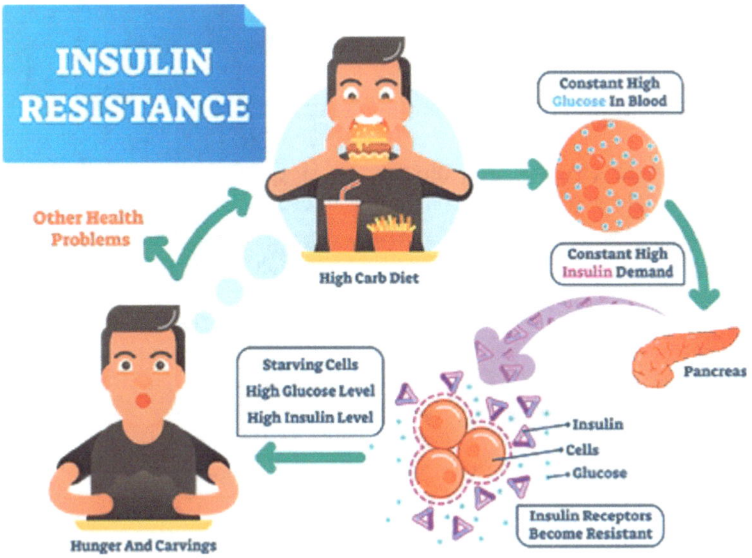

The trick here is to simply keep your insulin levels low by learning to eat high protein, low sugar (low glycemic), and possibly some high fat meals (only healthy fats) and steer clear of these sugary type meals.

- You'll still be able to enjoy yourself with fun foods, but whilst you want to burn fat, <u>you need to stick to the plan this section will outline for you</u>. There is no cheating, bioscience doesn't care about your feelings, so be disciplined, you'll need discipline, stay the course.

But first, what exactly is sugar? Is it just a white crystalline powder that we put in our coffee and tea? Afraid not readers.

Sugar - When I talk about sugar, I am not just talking about a white crystalline powder you have with your tea or coffee, I am also talking about pasta, breakfast cereals, rice, bread, potatoes, chips and many other simple carbohydrates that are laden with glucose. These food groups are your enemy; they hold little nutrients, contain huge amounts of energy and spike your insulin levels which tell your body to lock them away as fat.

Our bodies did not evolve to eat large quantities of these, however these foods are found almost everywhere nowadays as we are bombarded with (what I consider) poisonous simple carbohydrates.

Sugar is just a name given to sweet soluble carbohydrates. Names such as glucose, sucrose, lactose, maltose, fructose, galactose and more. Chances are that if it ends in 'ose' it will probably make you fat (even though some foods contain no fat).

And just to confuse you even more food companies disguise their added sugar in their ingredients list with names like nectar, cane crystals, dextrose, high fructose corn syrup, molasses, organic evaporated cane juice, crystalline fructose, fruit juice concentrates and more. Simply stop eating processed or packaged foods and you'll do fine in avoiding all these disgusting fake foods.

Stay clear of fruit juices to, they are packed with sugar even when it says "no added sugar", but we will get to fruit juices later.

Ketosis Mode Versus Glycolysis Mode

Glycolysis mode is when your body uses sugars in your blood for energy (you can't lose weight with this mode). Glycolysis mode is also when your body stores sugars, fats and carbs that you have eaten as fat. Again, high insulin levels from eating the wrong foods is what causes this.

Ketosis mode is when your body uses its fat reserves for energy (this is what you want). This mode is induced by your glucagon hormone.

Okay, so you want to be in ketosis mode (fat burning mode). But how do I get my body into ketosis mode? And once I am there, how do I stay in this mode?

We are going to learn how to eat and drink properly. I am going to outline a diet in the coming pages that you should follow quite stringently if you want to look good, feel good and be healthier.

But before we see that diet there are a few more rules to the game.

When we don't consume sugars, and instead eat lean foods such as salads or healthy fats, we force our body to use its fat reserves for energy. This is known as ketosis mode or fat burning mode and is where you want to be.

Carbohydrates (breakfast cereals, potato, rice, pasta, bread, sugar) = sugar = if unused turns to fat.

- Be very aware of foods that say zero carbohydrates or zero fat, because they still may contain huge amounts of sugar which get stored as fat once you eat them. Alternatively, foods that say no sugar may contain huge amounts of carbohydrates or fat.

Eat foods that that contain no sugar, no fat, no carbohydrates. We're looking for high protein / leafy green / healthy fats in our diet.

Does a banana contain sugar? Does a potato contain fat? The answer is no for both of those questions, however these are terrible foods to eat if you want to lose fat. These foods are packed with energy, will heighten your insulin level which takes your body out of fat burning mode and into fat storage mode.

Steer clear of these types of foods whilst you want to lose fat. You can eat them later (nothing unhealthy about bananas or potatoes), but when you are looking good and feeling good, chances are that you'll probably un-acquire the taste for some of them.

Two modes you will learn about here are "weight loss mode" (the hardest mode) and "maintenance mode" (keeping it off).

The weight loss mode can be hard, but keeping it off (maintenance mode) is easy with these techniques.

- The following is not a diet; it is a lifestyle that you will adapt into over a few months.

It may take some practice, but you'll get there.

Some rules to this game - Sugar Free Drinks

Even sugar free drinks will stop your fat burning process and raise your insulin levels simply by the chemical receptors on your tongue thinking you are consuming sugar and signaling your insulin levels to jump. **There is no cheating in weight loss (I wish there was).**

However sugar free or low carb, is always better than sugar.

The liver is a fascinating organ that performs over 500 functions inside the human body. However, the liver can only do one thing at a time (and one of its functions is fat burning). So apart from the occasional black coffee try and drink only water; soda water can be a nice change from standard water, but drink only water when you can (it allows the liver to slowly work your fat away).

Snacking in between meals is also a bad idea. After a meal, let your insulin levels drop down before your next meal. You can't lose weight whilst your insulin levels are elevated, however you can store fat whilst your insulin levels are elevated which is the opposite of what you want to do, which is probably why you are reading this section. No snacking after a meal, let your insulin levels drop back to baseline.

Every time you snack between meals you are losing valuable fat burning time and adding to the amount of calories your body may store as fat.

- So drop your insulin levels and stop snacking in between meals!

Retrain your taste buds to not consume sugar (they will adjust and you'll love the fact they did).
It may take a few weeks to retrain those taste spuds because you are probably accustomed to sugary foods, and this is probably what led you to this section, your weight. Use the 5% rule, just remove small amounts of sugar or carbs from your diet each week, until after six months (think of the math), you have removed over 100% of your sugary diet (bread, pasta, ice cream, cereals, rice, cake, etc).

Before I hear you say that you can't drink coffee or tea without sugar, you need to keep in mind that for millennia humans didn't have all the nice sugary foods we eat today. So you certainly can learn to eat and drink without sugar in your diet. Simply harden yourself up as you enter this journey. Your will power will increase your ability to taste normal foods without it being saturated in sugar and fats.

You will find that after a month, something amazing happens, your taste spuds become hypersensitive to sugars. A slice of bread can leave a sickly sweet taste in your mouth after eating it, because all of a sudden your taste spuds can actually taste how much sugar is put into these foods. This sickly sweet aftertaste that you will begin to notice in many foods you used to eat becomes your friend, because not only does it remind you about how sugar laden these foods are but you simply won't enjoy the taste anymore.

For every 1 gram of carbohydrates you eat (we're talking about potato, rice, bread, pasta and breakfast cereals), your body needs to retain 2.8 grams of water. Most people's stomach, chin, bum and thy bloating is simply water retention. Go on a super low sugar diet (salads) and you'll easily lose 5 kilo's in a fortnight (it's all water, not fat), but you'll love the weight loss either way.

Black coffee, black tea, green tea with NO SUGAR, (you'll get used to it, and once you do, you'll never go back to sugar). Of course you can go back to your wine and beer and fun foods later but whilst you are working on losing those kilos we'll stay disciplined.

Cut out late night snacks, learn to drink green tea or water at night time (your taste spuds will adapt and so will your resilience to snacking and junk food). You will become tougher and smarter as time goes on.

If you live with someone, it is much easier to lose weight if others are not eating junk food around you. Ask them to stop putting junk in front of you. Ask them to support you and to make healthy food (very lean protein with nice veggies such as onions, mushrooms, sunflower or pine nut seeds, leafy greens, tomato, etc).

Go to the shops, get smart and get creative. **Steer clear of processed foods!**

Learn to read the labels on the food packaging at shops. Also check the fat, sugar and carbohydrate content (make sure they are close to zero).

Your liver can only do one thing at a time, remember? If you want to lose weight you must be drinking water during the day or unsweetened black coffee or teas. No need to drink more than your thirst for.

Your body has a carbohydrate battery, that is, glycolysis mode. This is when sugars in the blood are responsible for energy. This mode is no good to you. You want to access your fat reserves used for energy, this is ketosis mode. You can only reach ketosis mode by eliminating the sugar in your blood (your carbohydrate battery), which means no sugar in your diet. Once your carbohydrate battery is depleted you can enter ketosis, where your body begins to burn stored fat. It's really that simple.

Remember sugar comes in the form of rice, all breakfast cereals, bread, pasta, potato, many fruits, cakes, chocolate (of course) and so on. I keep stressing this because we are conditioned to think these foods are normal.

So what should you eat? We will cover the diet for both ketosis mode and maintenance mode later.

But a brief overview could be foods such as lean fish (not from fast food take aways with chips and sugary drinks), or two scrambled eggs is about what a grown man (who trains) should eat for breakfast with a black coffee. There is no need for bread, bacon, hash browns, etc. Otherwise sliced and lightly fried smoked salmon is super healthy and a great bacon substitute. You won't miss bacon (or its poisons). However learn to skip breakfast and fast till lunchtime, combined with a low sugar diet and some quick high intensity training, you will be on the path to weight loss in no time at all.

Never eat until you are full, this will also slow down your metabolism.

A great time to train is when you first wake up and after a black coffee (on an empty stomach). Please don't add sugar (artificial sweeteners) or milk to your coffee, you'll get used to the taste.

You'll notice that most professional athletes train at 5a.m. This is because your body has not eaten for 7 to 8 hours and you are naturally in ketosis mode already (your insulin levels have dropped overnight and the glucose in your blood are at a minimum, so accessing fat reserves is easy).

Black coffee in the morning will keep you in ketosis mode and enhance cellular metabolism.

BTW, no need to train at 5a.m, just when you wake up, anytime is fine. Again, don't eat breakfast.

Don't cheat with sugar in your coffee or a biscuit here and there, you will be wasting your time, biochemistry doesn't care about your cravings.

Dieting is easier when your fridge has the right foods. Go to your local super market, read the labels of high protein foods you like. Make sure they are low in fat and carbohydrates. Fish, kippers (never in oil) other types of sea food, smoked salmon (tastes great lightly fried, good bacon substitute) white lean meats such as pork and chicken, maybe some red meat. Also cut the fat off these foods prior to cooking them. Don't eat sausages or meats that are processed.

Only spray a light film of oil across any cooking surface. If you tilt the pan sideways and oil pools in the corner you are using too much oil. Oil is a pure fat and something you don't want to be eating during weight loss mode (or ever).

Condiments like tomato sauce or mayonnaise are packed with sugar.

- **You may as well sprinkle sugar over your fish or chicken if you enjoy things like ketchup**.

So try not to use sauces and other condiments. Experiment with spices and herbs to spice up your foods.

Also go to your local health shop and purchase some protein powder with zero carbs, sugar and fat. Try and get one with no artificial sweeteners if possible, we want you to stay in ketosis mode. Chat with your local health shop about protein rich foods, they'll sell you nuts and other stuff you can use as a meal. Don't let them sell you those high protein bars with artificial sweeteners, they will pull you out of ketosis mode and you'll be wasting your time. Note – Health shops are there to make money and sell you products, they will try and sell you everything if they can.

Note, there is no such thing as a weight loss friendly breakfast cereal, it really is all complete B.S.

Read the content label, it's all just sugar (carbohydrates = sugar). The cereal you should consume (not for weight loss mode either, only whilst in maintenance mode) is a small bowl of porridge or oats (but only the porridge that contains no other ingredients except oats). Most oats or porridge are packed with salt, sugar and god knows what else, so read the ingredients, if it says anything else except oats, bin it.

Proper porridge / oats is okay to consume during maintenance mode because it is a slow release food but for weight loss mode you don't want to eat porridge, only when you get to maintenance mode. Try and live on protein, some veggies, water and black coffee. If you can then the water retention and fat cells will fall off you, especially if you do some high intensity interval training at the same time.

The more lean muscle you carry the more fat your body will burn as well. Try some light weights in the morning if you can, otherwise sprinting, pushups, jump squats to plough through your carbohydrate battery is great, otherwise **brisk** walking for 45 minutes is fine (just takes much longer). The harder you walk the more you burn. Wear earphones and listen to fast paced music that you love whilst you walk, music makes a huge difference (it's uplifting and makes you walk harder and faster).

Will power is like a muscle, the more you do this the easier it becomes until it is second nature. The first few weeks or months may be the hardest, but you'll become an expert at it after a while.

But depending on your discipline level, in 6 weeks to 6 months you'll start to be blown away by your results. Don't set unreasonable goals to look good, it can take time depending on your size. Weigh yourself today, then every morning. If you do the above you'll lose a minimum of 5 kilos in the first week. I've done this many times, and from experience it gets easier every time.

The above information is not a diet, it is simply how humans should live all the time. You really won't miss the chips and chocolate either (however once you are in maintenance mode you can indulge in bad food from time to time, you know I do).

There is a lot to absorb above, start making the gradual changes with the above (however the faster the better).

Losing weight is 98% diet. Personal trainers don't tell you this, they make money when you have a weight problem. Induce ketosis mode and weight loss happens. Never eat a meal larger than your fist. Two serves of food per day is fine, three if you are training. Never eat a big meal at night either. Try not to eat after 6.30pm.

If you do eat a pasta or rice dish (and we all do from time to time), try and eat those earlier in the day. Then consume the lean protein or salads later in the day.

Plain or lightly salted popcorn is very light in calories and so is beef jerky, so these can be good snacks if you are desperate for something quick.

Buy some no carb fibre supplements as well. High protein can slow your movements down, so include some fibre in your coffees and teas.

Some Interesting notes on burning calories. Study participants that slept in rooms cooled down to 18 – 19 degrees Celsius (66F), doubled the amount of brown adipose tissue they had, which is a type of fat that burns calories. This could be why many believe that when you feel cold you are burning energy / calories. So enjoy the cold outdoors and don't reach for that heater if you are trying to lose weight, make your body burn fat for the warmth.

In addition, with High Intensity Interval Training (HIIT), it was shown that woman performing 3 x 20 minutes HIIT workouts per week would lose almost 3kg extra per week than women doing standard exercise! HIIT works. And HIIT means greater consumption of oxygen post workout, so you are burning calories for a duration afterwards as well.

Snacking between meals is extremely detrimental if you are trying to lose weight. Snacking prevents your insulin levels from dropping which in turn keeps you out of fat burning mode. You cannot cheat me or others with a fat burning diet, only yourself.

Don't plan social engagements during this diet. No birthdays, no social engagements, no big sports games on with your friends coming around. Find a block of time where you can have no temptations for a few days (at least five days). Ask people close to you to not bring food over or invite you out whilst you are new to this diet.

Remove all the fun foods from your cupboards (chips, biscuits, sweets, etc). Don't eat them, throw them out or give them away. If you really need to have a snack, just a small handful of popcorn or a cup of extremely low calorie soup will do.

Use a standing up desk. If you sit down all day you will obviously burn less fat. Standing up keeps your leg muscles working and burns calories and does make a difference. So stop sitting for 12 hours a day (at work and at home) and learn to stand up whilst you are working.

Weight Loss Mode Diet and Regime

Your journey begins on this page, and this is the hardest part. Some of you will find it easy, others may not. But if you can get through this page and repeat for three days, everything else will seem easy afterwards.

You wake up. Black coffee, no milk, no sugar (you'll get used to the taste, so please harden up, its not the end of the world). You can have coffee with your lunch as well but try and cap it at three coffees per day.

90 seconds of extreme exercise (jumps squats, push ups, chin ups, sprinting). Choose only one of those exercises, whichever suits you. Yes, and only 90 seconds, but you better work hard otherwise you are wasting your time. Sprinting is a great starter, so do 3 by 30 second bursts of sprinting as fast as you can.

Walking and standing is great throughout the day, and more exercise is okay if you can handle the low sugar levels that you will be experiencing. This diet may rock you, seriously, you may feel light headed, you are guaranteed to feel lethargic, so be careful how much exercise you start off with. However these feelings are just side effects of burning fat, so be excited when you feel them.

No need for breakfast, your black coffee is enough.

Lunch - 200 grams of lean foods such as chicken, fish or vegetable protein (or a slow release carb). Make sure you are not eating any fats or batter with these foods otherwise they will take you out of fat burning mode.

Dinner - salads, vegetable protein or something similar. 200 grams in total.

You can drink as many herbal teas as you want during this day as long as they have no added sugars or sweeteners. Remember you can only cheat yourself, so check the boxes before drinking them, otherwise your efforts are for nothing.

Soda water (with nothing added) is also a good quencher with zero calories that can give your stomach the illusion of feeling full.

You need to repeat this diet for three days. It won't be easy, but it means you would have burnt through your carbohydrate battery (the stored sugars floating around in your blood). It also means that your body is now experiencing glycolysis mode (fat burning mode) and all your energy is coming from your stored fat reserves.

But not eating sugars or carbohydrates your body no longer needs to retain so much fluid and you will weigh anywhere from 2 to 5 kilos less within days, which in turn makes your face, stomach and butt to appear less bloated (the effects of low water retention are fast and a great motivator).

Maintenance Mode Diet and Regime (keeping it off)

Your new lifestyle begins on this page, and this becomes easier and easier as time goes on.

You wake up. Black coffee, no milk, no sugar (you'll get used to the taste, so again, harden up marine).

Some quick brisk walking for 45 minutes, or 90 seconds of extreme exercise (jumps squats, pushups, chin ups, sprinting). You need to choose only one of those exercises, whichever suits you, but you are welcome to do as much exercise as you can.

Work hard otherwise you are wasting your time. Sprinting is a great starter, so do 3 by 30 second bursts of sprinting as fast as you can.

Walking and standing during the day is great, though more exercise is okay if you can handle the low sugar levels (fatigue) that you will be experiencing.

There is no breakfast in flat stomach or small butt world.

Lunch, 200 grams of protein. Chicken, fish or meat or vegetable protein. Make sure you are not eating any fats or batter with these foods otherwise they will take you out of fat burning mode.

Dinner, more salads or protein. Try 100 grams of lean protein with a big salad.

Again, you can drink as many herbal teas as you want during this day as long as they have no added sugars or sweeteners.

Welcome to your new lifestyle. Think this is strict, boring or difficult? It becomes easier over time. And of course, you will gorge yourself from time to time, but other times you will maintain some semblance of diet and HIIT, which will ensure you can look and feel good.

Popcorn is another great snack. Cook your own and do not put sugar, honey or anything sweet onto them. Try sprinkling a small amount of cinnamon onto them maybe.

Remember, that earlier in this book we talked about Acetic Acid which slows down the absorption of sugars into the blood and also alleviates hunger. Apple Cider Vinegar contains acetic acid, and I occasionally use capsules if I am snacking or having an off day.

Don't use condiments such as tomato ketchup or mayonnaise and other similar products. You may as well sprinkle sugar on everything you eat. Pepper and some spices are okay so learn to use those instead.

Some Handy Combinations to Stay Healthy

Many foods are boring by themselves; however nature hasn't completely flunked in making healthy foods taste great. There are some great combinations that when paired become healthier than standalone foods. If you know what foods to pair you can enhance their flavor and reduce the amount of harm those foods do at the same time.

Red Wine and Fish - If you drink a little red wine with your fish it seems to elevate the levels of omega-3s in your body. Polyphenols which are strong antioxidants are contained in wine and suspected of causing this handy effect.1

Barbecued Meat and Dark Beer - Surely a match made in heaven. Even though I would steer clear of barbecued meat for longevity purposes, it is worth knowing. Marinade your meat in dark beer prior to cooking and the beer will reduce the intake of polycyclic aromatic hydrocarbons (PAH's) which can cause colorectal cancer among other nasty things. The darker the beer the better. 2

Cinnamon and Sugar - Cinnamon tastes great but has a secret weapon in its chemical makeup. Cinnamon contains a compound that assists your body to metabolize sugar, basically lowering your blood sugar levels. So when eating something sweet put some cinnamon onto it and help your body metabolize those sugars faster.3

Tomatoes and Olive Oil - Who would have thought that adding olive oil to your tomato soup would increase the amount of lycopene that your body can absorb. Lycopene is a great anti-cancer compound and cooking your tomatoes increases the amount of lycopene bioavailability as well.4

Long Pepper - Sprinkle some long pepper on your food and the active ingredient known as Piperine increases the bioavailability of the food. Use long pepper on your healthy soups and salads to increase nutrient absorption.

Green Tea with Lemon - Green tea contains catechins which are a natural phenol and antioxidant. Adding lemon juice to your green tea enhances the absorption of these catechins and tastes great.6

Avocados and Burgers. - Interestingly those of you that eat burgers with avocado have better circulation and less inflammation than those of you eating burgers without avocado. Avocado also contains healthy fats and tastes great.7

Grilled Meat (Red) and Rosemary - We all know Rosemary tastes great but research shows it reduces the amount of carcinogens known as heterocyclic amines which are produced during the grilling process. Add some Rosemary to those ribs next time.8

References :
1 Am J Clin Nutr. 2009 Jan;89(1):354-62.
2 J Agric Food Chem 2014 Mar;62(12):2638-43
3 Diabet Med. 2010 Oct;27(10):1159-67.
4 AmHeart J 2008 Jan;155(1):175-81
5 Asia Pac J Clin Nutr. 2005;14(2):131-6.
6 Mol Nutr Food Res. 2007 Sep;51(9):1152-62 & ScientificWJ 2012 Sep;637953
7 Food Funct. 2013 Feb 26;4(3):384-91.
8 J Food Sci. 2010 Mar;75(2):T40-7.

Food Porn – How the Author Eats

I will start my day fasting, and for lunch may have something that is extremely low in carbohydrates, fats and sugars. Here is an example -

Spinach, Strawberry, Walnuts and Flaxseed Oil Smoothie

Forget what the food industry is brainwashing us with. These foods are what we should be eating. If you want to **lose weight**, skip breakfast, make this smoothie at lunch time and watch it fill you up and your excess weight will fall off.

1. Wash your spinach (or other green leafy vegetable) even if the wrapping says that it is already washed.
2. Add three medium to large strawberries (or blueberries).
3. Add one tablespoon of flaxseed oil.

Put this into your favourite juicer and drink. Even though it is green and may look like a liquid vegetable, the strawberries and flaxseed oil give it a palatable creamy taste. The water content will fill you up, the lack of sugar (carbohydrates) in this drink will keep your insulin levels low which enables you to stay in fat burning mode from the night before.

The fasting effect is powerful due to you not eating since the night before and the fact that it is now lunchtime and it contains an extremely low amount of calories is fantastic with weight loss and anti-aging.

Preparation Time – 2 minutes

Ray's Anti-Aging Rating 10 / 10

Vegetarian Mince – Nachos – Taco's – Bolognaise

Using plant based protein (mince substitutes) it becomes very easy to make great meals that mimic mince based meals. I will go as far to say that most people would not even realize they were eating plant based mince unless they were told.

I always use a local plant based mince in Australia which uses myco-protein (from soy), but just do an internet search for your local meat substitutes (you may need to engineer these foods to suit your taste).

Recipe

1 packet of vegetarian mince substitute
2 tins of pureed tomato
1 packet of Mexican spice (experiment, many different types available).
2 cloves of garlic
1 dash of red wine (maybe)
1 dash of Olive oil

Put one table spoon of olive oil in a fry pan. Slice the garlic into tiny pieces and place into the oil and fry for 1 minute. Put both tins of diced tomato into the pan and continue to heat. Empty the vegetarian mince into the pan and break it up into the diced tomato (the two tins of tomato contain lycopene which when cooked becomes more bioavailable). Add a dash of red wine and the Mexican spice mix and stir until you have an even consistency.

Voila. Now you have a great tasting mince ready to serve into any mince based meal. Get some corn chips, jalapenos and some vegan cheese and Mexican food is ready to eat. Use this mince for any mince dish. Carnivores cannot tell the difference if you engineer the recipe well. Cooking time is less than 5 minutes.

Ray's Anti-Aging Rating – 8 of out 10 (but the recipe loses 2 points for the added wine).

Stir Fried Asparagus, Pine Nut and Pomegranate Seeds

An easy and great tasting stir fry which if cooked properly becomes quite addictive.

150 grams of Asparagus
150 grams of Green Beans
1 seeds from one Pomegranate
40 grams Pine Nuts
Salt and Pepper
Olive Oil
Butter

Optional Salmon or Potato or Tofu

Cut the base ends from your asparagus. Add four table spoons of olive oil to fry pan and heat. Fry the asparagus and green beans in a fry pan over high heat until the texture of the beans and asparagus is soft (cook to your taste). Whilst frying add very generous amounts of salt and pepper to the pan and a large dollop of butter (a golf ball will do).

Remove the asparagus and beans and place into a bowl. Throw the pine nuts into the fry pan and cook for a minute until they are lightly browned and then mix them in with the asparagus and beans.

Finally add the pomegranate seeds into the asparagus and bean mix and enjoy.

It is popular to add salmon, potato or other food groups into this meal.

Cooking Time - 10 Minutes

Ray's Anti-Aging Rating 9 / 10

Haloumi Salad with Almonds and Sultanas

Green Oak Lettuce
Red Onion
Cherry Tomatoes
Avocado
A Dash of Coconut Lime Dressing
Haloumi Cheese or a Vegan Cheese
With Lime and Pepper

Lightly fry the cheese. Mix these foods up to your specific preference. There is no right and wrong. Eat like this and lose weight and live long.

Be creative and add some options (basic food engineering). Options could include sultanas, apple, almonds and walnuts, etc. Get Creative.

Ray's Anti-Aging Rating 9 / 10 (we lose one point for the cheese)

Vegetarian Pizza

A great family friendly dish.

A pizza base (make it or buy it)
Pizza sauce
1 onion
100 grams mushrooms
Capsicum
Sweetcorn
Vegan Cheddar
Optional - Add Moroccan spice, pineapple, olives, etc.

You'll know what to do with the ingredients. It's easy eating healthy.

Ray's Anti-Aging Rating 5 / 10 (we lose five points for the bread base and cheese)

Soups!

Mushroom
Tomato
Coconut Curry
Pumpkin
Thai Red Curry Noodle Soup
French Onion
Mexican Lime Soup with Tofu
Spicy Asian Noodle Soup
And a hundred others,,,

Seriously, get creative! This is not a cooking book!

Every mouthful of food you have is either accumulating more cellular damage inside your body and making you age faster or removing cellular damage and extending your life. Choose carefully.

Ray's Anti-Aging Rating for soups is high, because they are mostly water and packed with the essentials of life.

High Intensity Interval Training (Only 3 Minutes Per Day!)

You can achieve the same biological advantages in three minutes per day, as someone that runs for 45 minutes. You won't have the same muscles as them, and hence won't be able to run for 45 minutes, but your cellular pathways will activate to purge free radicals, reduce blood glucose to bring you closer to fasting and weight loss mode and activate longevity pathways in the same way.

Jump Squats

When training, I want full range of motion from you – squat right down, jump straight up as high as you can. You may only be able to do five or six, do the rest of your squats without jumping. In a matter of days or weeks, you will have made vast improvements on how many you can do.

Start slowly if you are new to this type of exercise. Otherwise speak to your healthcare practitioner. Try aiming for 30.

Chin Ups

Remember, full range of motion. All the way down so your arms are straight, all the way up so your chin reaches above the bar. Palms front will give you a V shape, palms in will build your biceps.

These are tough exercises so go slow and ease into it. You can also use a pull down machine at the gym if your body weight is too much for your arms. However we are looking for exercises that leave you quite exhausted within sixty seconds. So none of these will be easy for first timers.

Triceps Dips

There are two ways you can do this. On a chair or park bench (for beginners). But work up to the free standing triceps dip bars (bottom photo) and learn to lift your entire body weight. 30 is your goal.

Full range of motion, all the way up, all the way down.

Pushups

As your resident immortalist, I want you to be able to drop and give me 40 (at least). Work your way up to this over a few weeks or months, no one can do 40 pushups without training the muscles first.

Keep your elbows out to the side as shown and make sure you achieve full range of movements. Arms fully straight then down to chest almost touching the ground. Anything else is not a pushup.

Pushups also engage your core, so ensure your torso is straight and your butt is not out of alignment. Also (yes I've seen people doing this) that your butt is not just moving up and down; the only thing that should move are your arms bending from straight to ninety degrees.

Sprinting

One of the best exercises you can do (if you can do it) is sprinting. One hundred meters or for thirty seconds.

Warm up first, possibly stretch if your healthcare practitioner advises it. Do three thirty second sprints or three one hundred meter sprints (uphill if possible).

Wait a few minutes between each sprint. At the end of every sprint, if you feel utterly exhausted, then you are kicking in your longevity pathways and on the right track (pun intended).

Do an online search for The Truth About Exercise. Of course you can also perform a similar exercise on a bike easily (go as hard as you can for thirty seconds).

Sit Ups – 4 Minutes – Tabata

Tabata is a form of interval training. I want you to find a four minute abdominal workout that leaves you wondering how you are ever going to get through the next three minutes once you have done sixty seconds.

So with this type of interval training (Tabata), you will perform an abdominal exercise for twenty seconds, and then rest for ten seconds before starting another abdominal exercise.

Find a video of a four minute abdominal workout you like, watch it a few times. Make strong mental notes of where the instructor's feet are (are they off the ground), how high the instructor sits up, do they completely relax when they uncurl back to the ground or do they leave their back slightly up, etc.

I reassert on every page in this section about full range or movement. What I am actually doing is making sure you understand technique. You can watch someone in the gym arm curl twenty kilo dumbbells, but once you show them the correct technique, that same person will struggle with ten kilos. Technique is what delivers results, not the weight or the speed you perform at.

Tabata ab workouts are great. I do them about two or three times per week. If you are new to this, then do not expect to make it through the four minutes. It is okay to lean into this over a period of weeks. But you will notice improvements every time you perform the micro four minute workout.

Find an online video of a four minute ab workout. Save it to your favourites (on your phone or wherever you train) and have it ready for whenever you need it. If there is music and the instructor is quite verbal, it certainly helps. A Tabata timer that you can also hear helps.

Softer Exercises for Older People

You might be out of shape, have an injury or other ailment that prevents you from lifting your body weight or accelerating your heart rate to quickly.

There are many exercises you can still perform, from riding bikes, walking to light jogging, to standing squats, etc. I cannot stress enough that moving your body is currently still one of the biggest weapons you have in longevity.

I am not judging people that use electric wheelchairs of course, many do not have a choice. But others use them because their weight gives them shortness of breath when walking. Sitting is the new smoking. Do everything you can to ensure you move.

Some elderly people may struggle getting up off a chair or into a car. Guess what, those are great exercises for elderly people. You might think that it won't prolong a ninety year olds life just by getting up out of a chair five times in a row for exercise purposes. I guarantee something as simple as that will extend life. So do not give up on your mum and dad's just yet, that tiny bit of exercise they do still activate all manner of cellular and biological pathways to keep things functioning well.

I totally encourage the elders among us to attend gym and lift weights and perform other exercises. You would have seen my colleague Dr Jeff Life earlier in this book. Jeff was 81 in those photos and can outperform most thirty year olds. Staying youthful truly is within our grasp.

Of course many elders may think of lifting weights, longevity, age reversal as silly, because it may sound so supernatural to them, but times are changing. Humans have already doubled their life spans in the last 120 years, and our technologies now are light years ahead of where they were in the 1950's. I contend that we are overdue for an explosion in human life span.

Older people in your family may exhibit muscle loss (sarcopenia) or struggle for basic movement. Keep them active for as long as you can, because cutting edge rejuvenation technologies are just around the corner.

Index

E

F

G

H

I

J